Water & the Shaping of CALIFORNIA

A literary, political and technological perspective on the power of water, and how the effort to control it has transformed the state.

by Sue McClurg
Foreword by Kevin Starr

Published by the Water Education Foundation and Heyday Books

We dedicate this book to California's past, present and future generations.

Cover photos from *Memoir and Maps of California, 1851*, courtesy of The Bancroft Library, and the California Department of Water Resources

Individual photo credits appear on page 164.

Acknowledgment for permission to reprint materials appears on page 163.

© 2000 by the Water Education Foundation

Published by the Water Education Foundation and Heyday Books, with major support from The Hans and Margaret Doe Charitable Trust.

Library of Congress Cataloging-in-Publication Data

McClurg, Sue
Water and the Shaping of California / by Sue McClurg ; foreword by Kevin Starr.
    p. cm.
Includes bibliographical references.
    ISBN 1-890771-37-6 — ISBN 1-890771-33-3 (trade paper)
    1. Water-supply—California-Management—History. 2. Water-supply development—California—History. 3. Water-supply engineering—California-History. 4. California—History. I. Title.
TD224.C3 MG3 2000
333.91'009794—dc21   00-09145

ISBN 1-890771-33-3 (Paperback)
ISBN 1-890771-37-6 (Hardcover)

Printed in the United States of America on recycled paper.

10 9 8 7 6 5 4 3 2 1

Editor: Rita Schmidt Sudman

Designer: Curtis Leipold

Literature Research: S. Joshua Newcom
Valerie Holcomb

Photo Research: Christine Schmidt

Printing: Paul Baker Printing

WATER EDUCATION FOUNDATION
717 K Street, Suite 317
Sacramento, CA 95814
(916) 444-6240, fax (916) 448-7699
www.water-ed.org

Individuals may purchase copies from:

Distributed to the trade by:

Heyday Books
P.O. Box 9145
Berkeley, CA 94709
(510) 549-3564, fax (510) 549-1889
www.heydaybooks.com

Acknowledgments

For several years the Water Education Foundation's chief writer, Sue McClurg, and I worked on this book that chronicles the story of California's water and the relationship between that water and the people of California. It's our mission as a non-partisan Foundation to create a better understanding of water issues and help resolve water resource problems through educational programs. This book represents our biggest effort toward that goal.

Several people helped us create this book. The book began with a gift of photos of California taken during the 1987-1992 drought by *National Geographic* photographer Rick Rickman. Rick traveled across California in search of images illustrating the intense drought and how people were coping with it. Along the way, he became very involved with his subject. Naturally,

only a handful of his great photos were published in the magazine. About forty became the starting point for this book.

Another man's passion for water helped as well when the Hans and Margaret Doe Charitable Trust gave us a grant. Hans was born in Norway and eventually made his way to the small southern California town of Vista where he became an avocado and macadamia nut rancher. His personal involvement in water issues influenced the delivery of water to Vista and all of southern California.

By the time Hans died in 1988 at the age of 85, he

Rick Rickman

and his wife had set up their trust which has since funded a number of educational water projects including public television documentaries, maps and school programs. I grew up in Vista and remember my mother often saying "God made the earth and Hans Doe made Vista." My parents were grateful for

Hans' efforts to secure a firm water supply and provide flood protection for our little town. While serving on the local water board, Hans also worked to resolve statewide water issues. In a 1978 interview with an area newspaper, he reiterated his philosophy, "God has given us the resources, and it's up to us to develop them." He always took any opportunity to talk about water reliability in an arid land and to remind the public of the old Ben Franklin adage, "When the well's dry, we know the worth of water."

Once the book was well under way we turned to two experts on California water history: Linda Vida, Librarian at the University of California's Water Resources Center Archives, provided invaluable help in locating historic photos and research materials. Former State Archivist John Burns reviewed the historical sections for accuracy.

Others also helped. In addition to Rick's photos, we used many images from the California Department of Water Resources. Charles Soderquist provided "Words for Moving Water" on page 160.

Experts who provided technical review included Ray Barsch, Joe Burns, Jason Douglas, Mike Eaton, Paul Engstrand, Brent Graham, Carl Hauge, Norris Hundley, LeVal Lund, Teddy Morse, Maurice Roos, Curt Spencer, Joyce Tokita and Gary Weatherford.

Water & the Shaping of California presents the facts about California's water and explores the politics of water. But this is not a book just about the facts and arguments. It is about the soul of water and what it does to our hearts. It's about the power of water and how it moves us: the beauty of water and how it gives us peace.

– Rita Schmidt Sudman
Water Education Foundation

Hans Doe

contents

Foreword

ater is at once a simple and a complex subject. Regarding water, we can make the most basic of statements: water is life itself. From water, arose in primordial times, all that now flourishes on the planet. Eons later, water remains the primary constituent and building block of creation. Thus the great religions of the world have each seen in water a primary symbol of the mystery of life.

Notice, however, how quickly water, taken at its most basic level, rapidly demands a more complex interpretation. A basic resource of the planet, water instantly becomes the progenitor and sustainer of all life. A simple natural element becomes the most powerful and universal of religious symbols.

Historians of civilization, moreover, have seen in the ancient irrigation projects of the Tigris-Euphrates Valley the beginnings of urban civilization. Cooperating in the pursuit and acquisition of water, human beings learned even further modes of cooperation; and from the skills they acquired in seeking water, they evolved the first cities, which soon engendered a wide array of arts and sciences. From this perspective, water is not only the basic fact of life and a primary symbol of creation, water is the mother of all civilized invention.

Nowhere is the primal and civilizing power of water more evident than in California. Through water, as this delightful book shows, Californians literally invented and shaped their commonwealth. Gold was important, true; but the Gold Rush yielded to an era of agriculture and urbanization that has lasted to this day – and that meant water, even more than it meant gold. In the arid and the semi-arid Far West, of which California is a part, water – the finding of water, the acquisition of water, its transfer and its usage – was the basic challenge of communal development. From this perspective, it can be said without exaggeration that American California across the past 150 years has literally invented itself through

water. The technology of the Gold Rush – which was a technology of land and water movement – became the technology of agriculture, also a work of land and water; and the technology of irrigation, in turn, became through the successive technology of reservoirs and aqueducts the technology of urbanization. Through water, in other words, mining led to agriculture, and agriculture led to irrigation, and irrigation led to an urban and suburban civilization based upon bringing water from where it was to where it was needed.

At once simple and complex, water is the result of a relatively simple arrangement of molecules; yet these arrangements contain within themselves the complexities of all natural life and, through human enterprise, the complexities of civilization itself. In its complex dimension, in fact – in the fields, that is, of water technology, water law and water politics – water soon becomes so complex that only the most advanced experts can understand any one or another phase of its operation. Hence the value of this book. *Water & the Shaping of California* invites us to consider how California – as environ-

Dr. Kevin Starr

ment and as human culture – has been shaped by the necessities of water. Like water itself, this book is at once simple and complex: simple in its direct presentation of fact and image, complex in the moral and imaginative message communicated by the basic story of water in California as how we human beings have employed water to create our civilization.

No one, finally, can be an aware Californian without an understanding of how and why water has played such an important role in the creation of the state. In water, we encounter the past, present, and future of our great commonwealth. This brief but pertinent book will help each of us fix in mind the essential facts and the revealing images of the California water story. Understanding water, we understand California; and in understanding California – including its creation through water – we understand even better something very important regarding the larger American experiment.

– Dr. Kevin Starr
State Librarian of California
Author, Americans and the
California Dream Series

Introduction

Water has the power to nurture, refresh, sustain life. Water is essential. We drink it. It grows our food, powers our turbines for electricity and serves as the lifeblood of our industries. It sustains our landscapes and provides habitat for fish and wildlife. Our lives and economy depend on having a reliable supply of clean, fresh water.

Water also has the power to destroy. Let loose from its banks, a river's destructive force can wipe out crops, wash away houses, even kill.

Water's absence also is powerful – when it dries up, it leaves land barren and uninhabitable. Its reappearance can be in such quantities as to be overwhelming.

In California, the natural power of water has been reshaped by engineering feats, political and judicial decisions and popular opinion. It is not a new story. Throughout time, from earliest recorded history, man has sought to control water, for he has always understood it to be a life-and-death resource. A look back through time finds evidence of this effort; ancient irrigation canals in the fertile crescent of Mesopotamia, the Roman aqueducts and Aztec terraces for farming are but a few examples.

As society has advanced in technological skills and innovation, the ability to control water and reshape our universe has increased.

Without the drive to harness the power of water and without the technological advancements of our society, the miners, farmers, city dwellers and engineers could never have fashioned the California we know today. The other forces for change – politics, the drive to settle the West and an ever-growing population – gave California extra incentive to create the most complex and sophisticated water delivery system in the world. The system, in turn, allowed the state to become the world's eighth-largest economy.

Yet our system has its problems. As we sought to control floods, store water for dry times, irrigate what was once desert, and use the power of water to generate electricity, we changed the very nature of the landscape. As society now questions the wisdom of such radical changes to the state's waterscape, a new era unfolds, one in which the developed system is being modified to benefit the natural system.

The ability to control water on a grand scale and the changed waterscape are only one part of the California water story. Water nourishes not only our bodies and our lives but our souls. The power of water to soothe the spirit, inspire the artist and encourage self-reflection can be found in the poems, essays and other literary sayings in this book. The selections themselves illustrate again this basic fact: water is a powerful force in our lives.

This book focuses on the power of water and California's efforts to use that power to sustain life, harness that power to improve life and control that power to protect life.

It begins with the natural California and its varied terrains – high mountain ranges to fertile valleys to arid deserts – and ends with the modern-day effort to restore some of the natural environment that has been altered by the changed waterscape. In between we explore the forces of nature – the great floods and the great droughts – that led, in part, to the construction of the great projects that created the California we know today – the nation's leading agricultural state and a magnet for an ever-growing urban population. A state in which the natural environment has secured its place in society, where not only heart and soul speak of its value, but also law and politics.

– *Sue McClurg*

A Natural State

California as it exists today is not the natural California. Through development of the state's great water projects, the liquid gold has been relocated from where it naturally occurs to where people wanted it sent.

"Physiographic California, for much of its length, is divided into three parts . . . : the Sierra Nevada, highest mountain range in the Lower Forty-eight; the Great Central Valley, essentially at sea level and very much flatter than Iowa or Kansas; and the Coast ranges, a marine medley, still ascending from the adjacent sea."

– John McPhee
Assembling California

The Story of California is the story of water.

It is a story of growth, growth that began with the Gold Rush, continued through the subsequent decades and remains a constant force in what is often referred to as the Golden State.

It is a story of economic develop-ment. From mining to farming to aerospace to high tech, California today ranks as the world's eighth-largest economy, a nation-state with $963 billion in total goods and services. It is a story of technology. From the hydraulic mining equipment of the 1800s to the electric turbine pump of the 1920s to the ozone water treatment plants of the 1980s.

Despite the state's alteration, its residents remain connected to that natural California. Drawn to the open space and wilderness of national and state parks, publicly owned ocean beaches and artificial and natural lakes, water retains its power to sustain the soul even as we rely on it to sustain life and the society we have built. •

Before the Spanish explorers, before the fur trappers, and before the '49ers, the territory that would become the nation's thirty-first state was home to Native Americans. Tribal groups were located throughout California speaking over 100 different languages and dialects.

Tribes in the Central Valley included the Maidu, Miwok, Yokut and Wintu. Coastal tribes, north to south, included the Yurok, Yuki, Pomo, Ohlone, Chumash and Gabrielino. Inhabiting the inland south state were the Serrano, Cahuilla and other Shoshonean tribes. Living in the foothills and mountains that form California's eastern border were the Washoe and Northern Paiute tribal groups.

While the tribes along the lower Colorado River did practice irrigation, the majority of these Native American groups' society was based on hunting and gathering. The most common archeological specimen in California is the stone grinding bowl or mortar, and the acorn served as the dietary staple for most groups.

Oak trees dotted much of the natural California landscape and, though time-consuming, the process of removing tannic acid and grinding acorns created a food rich in nutrients. Acorn pounded into a flour formed a kind of porridge to accompany fish, meat, berries, nuts or seeds.

Hunting and fishing were done mainly with nets and traps. Mountain tribes relied on the yearly upstream migration of chinook and coho salmon, using traps and sharpened poles to catch the salmon as they returned to spawn. •

California's Natural Waterscape

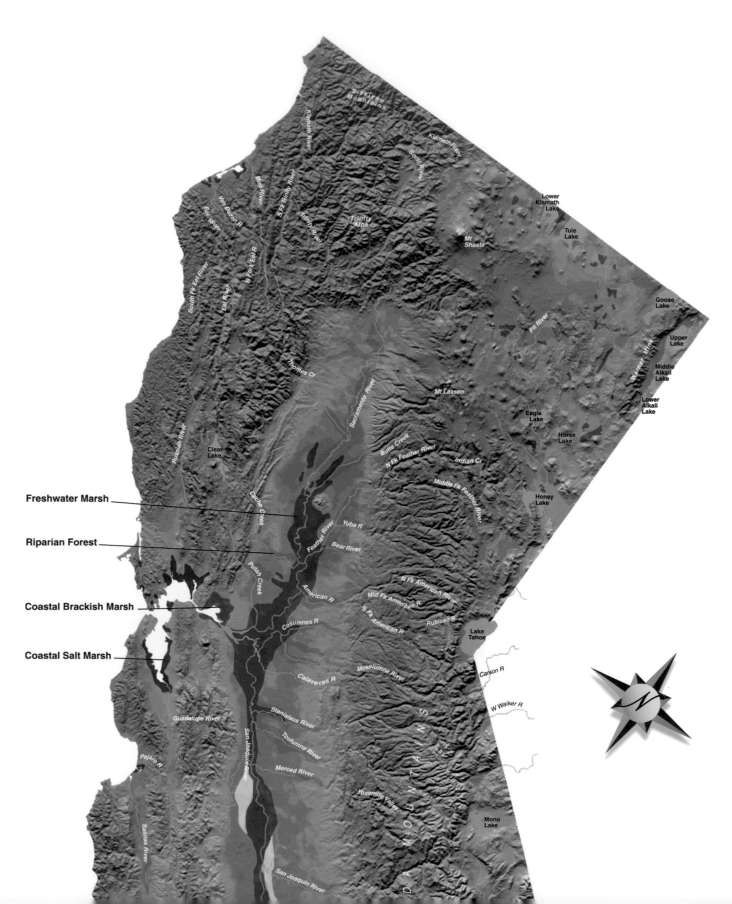

Freshwater Marsh

Riparian Forest

Coastal Brackish Marsh

Coastal Salt Marsh

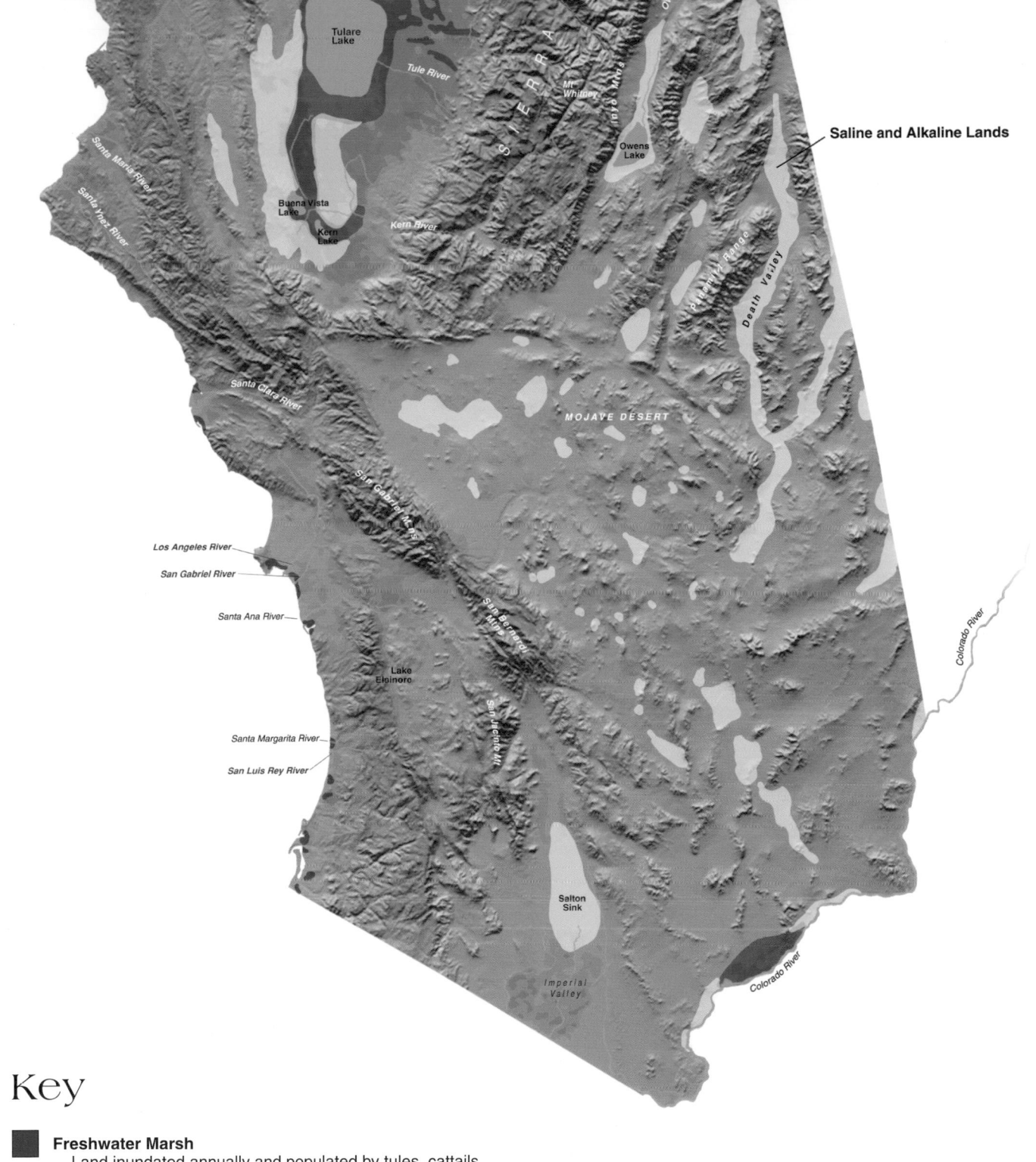

Saline and Alkaline Lands

Tulare Lake

Tule River

Mt Whitney

Owens Lake

SIERRA

Mt Whitney

Santa Maria River

Santa Ynez River

Buena Vista Lake

Kern Lake

Kern River

Panamint Range

Death Valley

Santa Clara River

MOJAVE DESERT

San Gabriel Mts

Los Angeles River

San Gabriel River

Santa Ana River

San Bernardo Mts

Lake Elsinore

Colorado River

Santa Margarita River

San Jacinto Mt

San Luis Rey River

Salton Sink

Imperial Valley

Colorado River

Key

■ **Freshwater Marsh**
Land inundated annually and populated by tules, cattails,
or other hydrophytic vegetation.

■ **Riparian Forest**
Broadleaf deciduous forest growing naturally on the sides
or banks of rivers and streams, and in bottomlands.

■ **Coastal Brackish Marsh**
Land inundated alternately by saline water and fresh water.

■ **Coastal Salt Marsh**
Land along the upper intertidal zone of protected shallow bays,
estuaries and coastal lagoons. Salt tolerant plants predominate.

■ **Saline and Alkaline Lands**
Sinks and basin rim lands characterized by intermittent water
high in mineral content.

Scale

| 0 | 10 | 20 | 30 | 40 | 50 | 60 | 70 | 80 mi. |

| 0 | 20 | 40 | 60 | 80 | 100 km. |

A landscape of contrasts between lush rain forests and dry deserts, granitic peaks and seaside beaches, flat valleys and steep cliffs comprises the natural state. Volcanoes, earth-quakes and water shaped these physical features of California.

In the north is the area of greatest precipitation, dominated by the redwood forests of the coast and the snowy peaks of the Klamath and Cascade mountain ranges farther inland.

The Cascades form the northern boundary of one of the state's most prominent features – the Great Central Valley. Bordered on the east by the Sierra Nevada and on the west by the Coast Range, the Valley stretches some 400 miles north to south and 50 miles east to west.

Over 120 million years ago, before the Sierra Nevada and Coast Range rose from the sea, the valley floor formed the bottom of an ancient ocean. As sediment eroded from the mountains, the valley gradually filled in. In its natural state, the valley was an oak savannah, home to pronghorn antelope, tule elk and even grizzly bears. The area at the confluence of the Sacramento and San Joaquin rivers, the Delta, was a marsh filled with reeds and tule rushes. Winter rains and rising rivers filled the valley with an inland sea.

The majestic Sierra Nevada was created by the raising of a great block of the earth's crust along a large fault on its eastern edge so all the rivers flow toward the west. At more than 400 miles, the Sierra Nevada is the longest mountain range in the United States. It features some fifty peaks more than 13,000 feet in elevation, including America's tallest mountain outside Alaska. At the western edge of the Sierra Nevada are the Central Valley foothills, famous for their gold ore and site of many historic mining camps.

The Tehachapis form the Central Valley's southern boundary. Annual precipitation south of the Tehachapis is much less than in the northern state. River flow here is intermittent in many places as the southern coastal plains give way to the dry Mojave and Colorado deserts.

Mt. Whitney at 14,496 feet marks the state's highest elevation. Nearby, but far, far below, is Death Valley – at 282 feet below sea level, it is the state's lowest point.

For early inhabitants and explorers, the formidable mountains and harsher environments formed natural bound-aries for settlements.

For the modern Californian, these geologic juxtapositions have been bridged by highways and motor vehicles, planes and trains – putting rain forests, mountain peaks, valleys, deserts and ocean within reach of a day's travel. ●

H istorically, more than 5 million acres of seasonal, permanent wetlands and riparian habitat once existed in California. Wetlands were found statewide from the Modoc Plateau and the Klamath River Basin in the north to the Imperial and Coachella valleys in the south.

At one time, less than 5 percent of these natural wetlands remained. Early state policies encouraged farmers to reclaim and plow swamplands while cities filled wetland areas to allow for economic development.

At the time of the Gold Rush, for example, San Francisco Bay had some 200,000 acres of tidal marsh. By 1965, only 35,000 acres of tidal marsh remained and the bay's open water had shrunk from 787 square miles to 548 square miles. ●

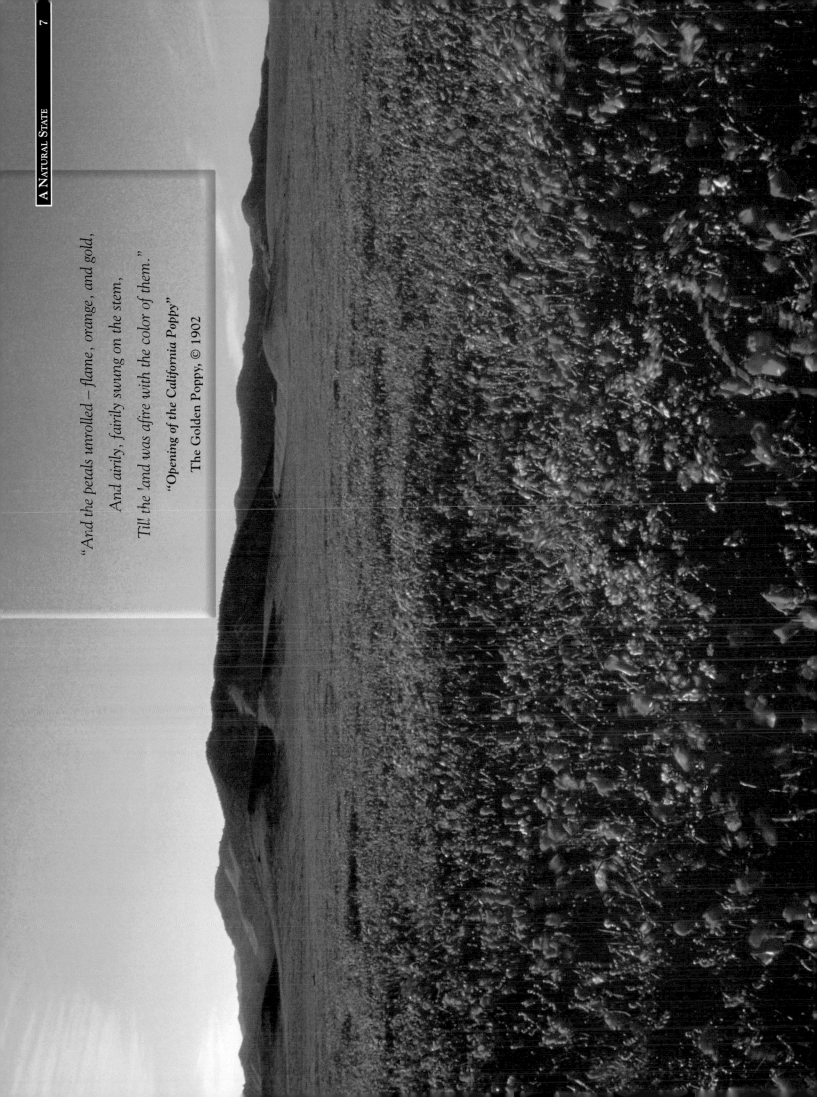

"And the petals unrolled – flame, orange, and gold,

And airily, fairily swung on the stem,

Till the land was afire with the color of them."

"Opening of the California Poppy"

The Golden Poppy, © 1902

"The white people go to the river and turn it into dry land.

The water says, 'I don't care. I am water.

You can use me all you wish. I am always the same.

I can't be used up. Use me. I am water. You can't hurt me.'

The white people use the water of sacred springs for their houses.

The water says, 'That is all right.

You can use me but you can't overcome me.'

All that is water says this,

'Wherever you put me I'll be in my home. I am awfully smart.

Lead me out of my springs, lead me out of my rivers,

but I came from the ocean and I shall go back into the ocean.

You can dig a ditch and put me in it,

but I go only so far and I am out of sight.

I am awfully smart.

When I am out of sight I am on my way home.'"

– *Wintu woman's tale, 1930* –
Natural World of the California Indians

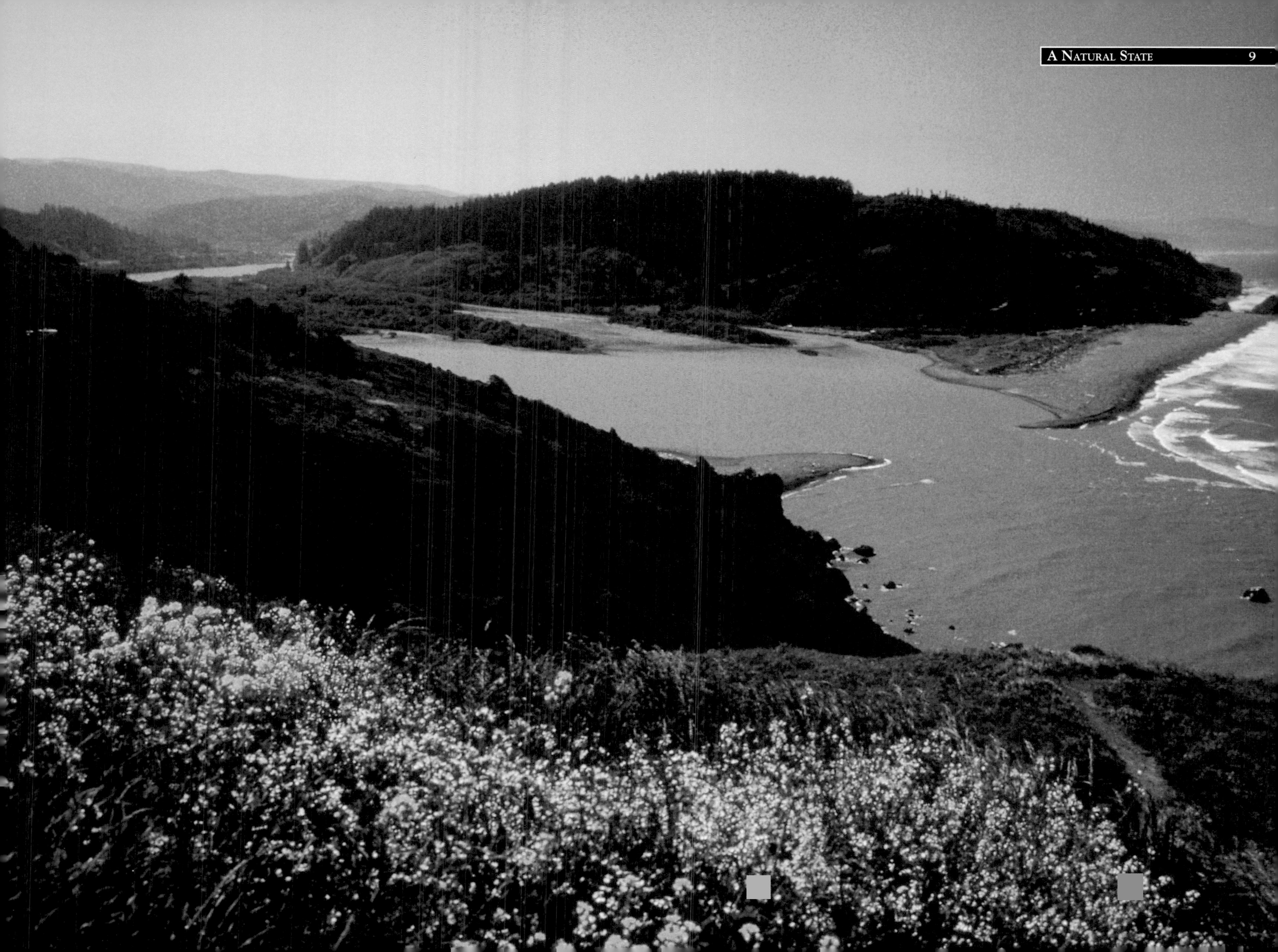

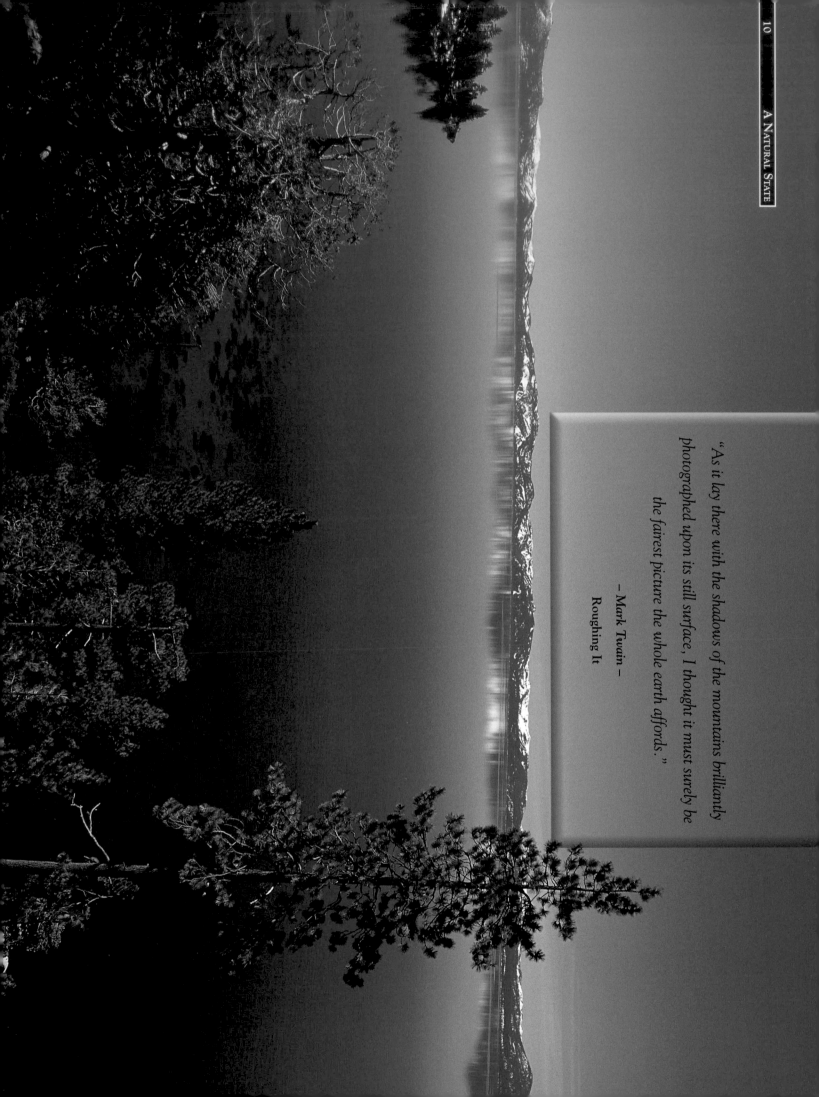

"As it lay there with the shadows of the mountains brilliantly
photographed upon its still surface, I thought it must surely be
the fairest picture the whole earth affords."

– Mark Twain –
Roughing It

"Our road was rough and difficult;

but, after traveling for three days we arrived upon

the brink of the great valley. The first view of this sublime scenery

was so impressive, that we were delayed for a long time,

as if spellbound, looking down from the mountain

upon the magnificent landscape far below. . . .

But, however grand the valley looked from above,

it was not until the next day, when we descended into it

and looked upward, that we obtained the grandest views;

just as, at Niagara, the most awe-inspiring sight is from

the foot of the falls, looking up at the waters,

pouring, as it were, out of heaven."

— James Capen "Grizzly" Adams —
The Adventures of James Capen Adams

"I have never seen Nature when she seemed so little 'Mother Nature'
as in this place of rocks and snow, echoes and emptiness."

– Clarence King –
Mountaineering in the Sierra Nevada

Nature's Reservoir

It is a cycle as old as nature itself. Locked in ice, billions of gallons of water are stored each winter. With the warmth of spring, nature releases water from this snowy reservoir, water to nourish the first green blades of new green grass, water to sustain life.

The thaw starts with just a trickle. As the days grow warmer, the trickle becomes a torrent. With the force of gravity water tumbles over boulders, drops off cliffs and thunders into streambeds. Rivulets become streams and streams become rivers as the fast-moving water rushes down the mountains toward the valleys below.

Most of California's residents live below the snow line. The area above the snow line offers a seasonal winter wonderland ideal for a weekend's visit without the inconvenience of shoveling the walk. California's natural climate leads to two distinct weather patterns: dry and wet, with roughly six months of each. Many other states, by contrast, experience rainfall throughout the year.

To stretch the water into a year-round supply, water projects were built to capture the snow melt of each spring and store it for use during the dry summer and fall. •

"Every tree during the progress of gentle storms is loaded with fairy bloom at the coldest and darkest time of year, bending the branches, and hushing every singing needle.

But as soon as the storm is over, and the sun shines, the snow at once begins to shift and settle and fall from the branches in miniature avalanches, and the white forest soon becomes green again."

– John Muir –
The Mountains of California

"Go as far as you dare in the heart of a lonely land,
you cannot go so far that life and death are not before you.
Painted lizards slip in and out of rock crevices,

and pant on the white hot sands.

Birds, hummingbirds even, nest in the cactus scrub;
woodpeckers befriend the demoniac yuccas; out of the stark,
treeless waste rings the music of the night-singing mockingbird.

There is neither poverty of soil nor species to account
for the sparseness of desert growth, but simply that each plant
requires more room. So much earth must be preempted
to extract so much moisture. The real struggle for existence,
the real brain of the plant, is underground;
above there is room for a rounded perfect growth.
In Death Valley, reputed the very core of desolation,
are nearly two hundred identified species.

The desert floras shame us with their
cheerful adaptations to the seasonal limitations.

Their whole duty is to flower and fruit, and they do it hardly or with tropical luxuriance, as the rain admits. It is recorded in the report of the Death Valley expedition that after a year of abundant rains, on the Colorado desert was found a specimen of Amaranthus ten feet high. A year later the same species in the same place matured in the drought at four inches. ... There are many areas in the desert where drinkable water lies within a few feet of the surface, indicated by the mesquite and the bunch grass (Sporobolus airoides). It is this nearness of unimagined help that makes the tragedy of desert deaths. It is related that the final breakdown of that hapless party that gave Death Valley its forbidding name occurred in a locality where shallow wells would have saved them. ... To underestimate one's thirst, to pass a given landmark to the right or left, to find a dry spring where one looked for running water — there is no help for any of these things."

— Mary Austin —
The Land of Little Rain

"California is gold-tan grasses, silver-gray tule fog, olive-green redwood, blue-gray chaparral, silver-hue serpentine hills. Blinding white granite, blue-black rock sea cliffs. Blue summer sky, chestnut brown slough water, steep purple city streets — hot cream towns. Many colors of the land, many colors of the skin."

– Gary Snyder –
A Place in Space

Early Life…

Mission San Diego de Alcala, founded in 1769 by Father Junipero Serra, was the first of twenty-one missions established by the Spanish in what was then known as Alta California. Water was a critical factor in the decision of where to locate each mission. In fact, the first mission San Diego had to be relocated down the bluff to a more favorable site on the San Diego River to gain a more reliable water supply.

As they made their first plantings of corn, beans and wheat, the mission padres realized they would need to irrigate these crops. The Spanish brought with them their system of water rights, which favored the right for everyone within a pueblo (town) to use water to satisfy their needs regardless of ownership of land.

These pueblo rights – furthered under the Mexican rule of California – later came into conflict with the English system of riparian water rights where rights were attached to land ownership, a system adopted when California became a state in 1850. •

"Wednesday, August 2 [1769]

We set out from the valley in the morning and followed the same plain in a westerly direction. After traveling about a league and a half through a pass between low hills, we entered a very spacious valley, well grown with cottonwoods and alders, among which ran a beautiful river from the north-northwest, and then, doubling the point of a steep hill, it went on afterwards to the south. Toward the north-northeast there is another river bed which forms a spacious water-course, but we found it dry. This bed unites with that of the river, giving a clear indication of great floods in the rainy season, for we saw that it had many trunks of trees on the banks. We halted not very far from the river, which we named Porciuncula. Here we felt three consecutive earthquakes in the afternoon and night. We must have traveled about three leagues to-day. This plain where the river runs is very extensive. It has good land for planting all kinds of grain and seeds, and is the most suitable site of all that we have seen for a mission, for it has all the requisites for a large settlement."

– Juan Crespi –

Describing the Los Angeles River and Plain

Fray Juan Crespi, Missionary Explorer on the Pacific Coast 1769-1774

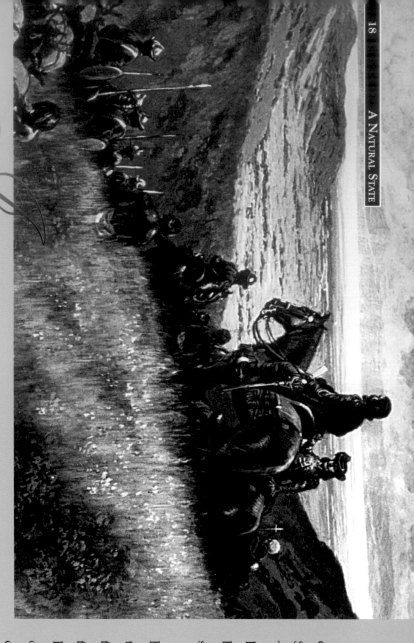

uan Rodriguez Cabrillo is believed to have been the first of the Spanish explorers to set foot within the boundary of California, landing in what is now San Diego during his voyage of 1542. Thirty-seven years later, England's Sir Francis Drake sailed his ship, the *Golden Hind*, into northern California, possibly landing in what is now known as Drake's Bay, north of San Francisco. Subsequent explorers mapped Monterey Bay to the south, but San Francisco Bay, the world's largest natural harbor, remained undiscovered for nearly two more centuries.

Even then San Francisco Bay was discovered not by sea but during an overland journey by members of Spanish Captain Gaspar de Portolá's 1769 expedition. Portolá's expedition had been assigned to establish a mission near Monterey Bay as the Spanish continued to expand their presence in California.

However, the explorers failed to recognize Monterey Bay from earlier descriptions and continued north, ending up camped in what is now San Pedro Valley on October 31, 1769. The next day, a Portolá scouting party led by Sgt. José Francisco Ortega became the first Europeans to catch sight of San Francisco Bay.

As Father Juan Crespi recorded in his diary, "As soon as we ascended to the summit, we descried a great bay formed by a point of land which runs far out into the open ocean and looks like an island. ... Following the coast of the bay to the north, some white cliffs are visible, and to the northwest is the mouth of an estuary which seems to penetrate into the land."

Although Crespi noted that, based on navigational information from earlier explorations, the Port of Monterey was behind them and that they had reached the port of "our Father San Francisco," the party actually thought they were at Drake's Bay, which other Spanish explorers had dubbed San Francisco Bay.

Other scouting parties continued to explore the area for several more days. Whether these parties ever caught sight of the Golden Gate – the narrow opening leading from the ocean into the Bay – remains in historic dispute.

What is known is that it wasn't until 1775 – seven years later – that the first Spanish ship, the *San Carlos*, sailed into San Francisco Bay.

The Spanish went on to establish the San Francisco de Asis Mission and Presidio in 1776, the sixth of twenty-one missions established between 1769 and 1823. Dotted along the coast, these outposts on El Camino Real (the King's Highway) formed a chain that legend says were no more than one day's journey apart.

Originally named Yerba Buena by the Spanish, the city was renamed San Francisco in 1847, a year after passing into control of the United States. Many of California's modern communities were built on the foundations of the early missions. Left as a legacy are their Spanish names – Santa Barbara, San Juan Capistrano, San Fernando, San Diego, San Luis Obispo, Santa Cruz, Santa Clara, San Jose and San Rafael. ●

The Romance of the Ranchos

n 1821 California became part of the new country of Mexico and the socioeconomic structure of the missions gradually gave way to the ranchos. Mexico expanded the Spanish tradition of making large land grants for ranchos, especially after the missions were secularized starting in 1834.

Spread along the coast from San Diego to Sonoma, some 800 families controlled the 10 million acres deeded to these new ranchos. The typical rancho dwelling was a long, one-story adobe with shaded verandah often with an interior courtyard. An individual rancho might encompass from 50,000 to 150,000 acres, comprised of several land grants.

The Spanish missions fell into disrepair as cattle ranching became predominant.

Cattle raised for their hides and tallow roamed over thousands of hills – commerce to be traded with British and American ship captains. Cattle ranching was the primary, and almost sole, business. Meals were beef for breakfast, beef for lunch, and beef for dinner, supplemented by grain, fruit and vegetables produced on the rancho. It was a time of romance and rodeos as vaqueros roamed the range. Children, poor relatives and even strangers might enjoy the hospitality of one of these ranchos, with Indians often used as servants.

Bull and bear fights in the pueblo and grizzly bear and elk hunts were occasions for both sport and work. The rodeo was a gala occasion, held not as a competition but as an opportunity to claim individual ownership of livestock on this unfenced range.

Mexican California remained isolated from the unrest in the Republic of Mexico and these early settlers, 6,000 non-Indian residents in 1840, did not explore much beyond the coast. There was plenty of land and resources to support the small population, and the coast provided some protection in such a sparsely settled land.

California's destiny was changed in the summer of 1846 when America declared war on Mexico. With the end of the war, a treaty of peace ceding California to the United States was signed on February 2, 1848 – just nine days after gold was discovered on the American River. •

ocial life on the ranchos was a continual round of festivals and fandangos, races and rodeos, weddings and wakes. Dances celebrated every life ritual.

Weddings were arranged affairs by fathers. Many such a marriage served to increase the size of the rancho as land holdings were combined. Women commonly were betrothed and married by fourteen or fifteen; early marriages were one method used to ensure their virginity.

When the fathers agreed upon a wedding day, they went to the church and informed the priest of this fact. Once the banns were published, the pair was blessed by the priest and proclaimed married before God and men. The ceremony itself was a solemn affair, preceded by confession and accompanied by no music.

Celebration followed as family and friends waited outside to escort the couple back to the house. The escort often included a band on horseback, and horses and wagons were festooned with bright handkerchiefs and flowers. At home there was a dance that went on for at least two or three days, sometimes as long as eight, and every day a calf or two was killed to feed the hundred or so guests. •

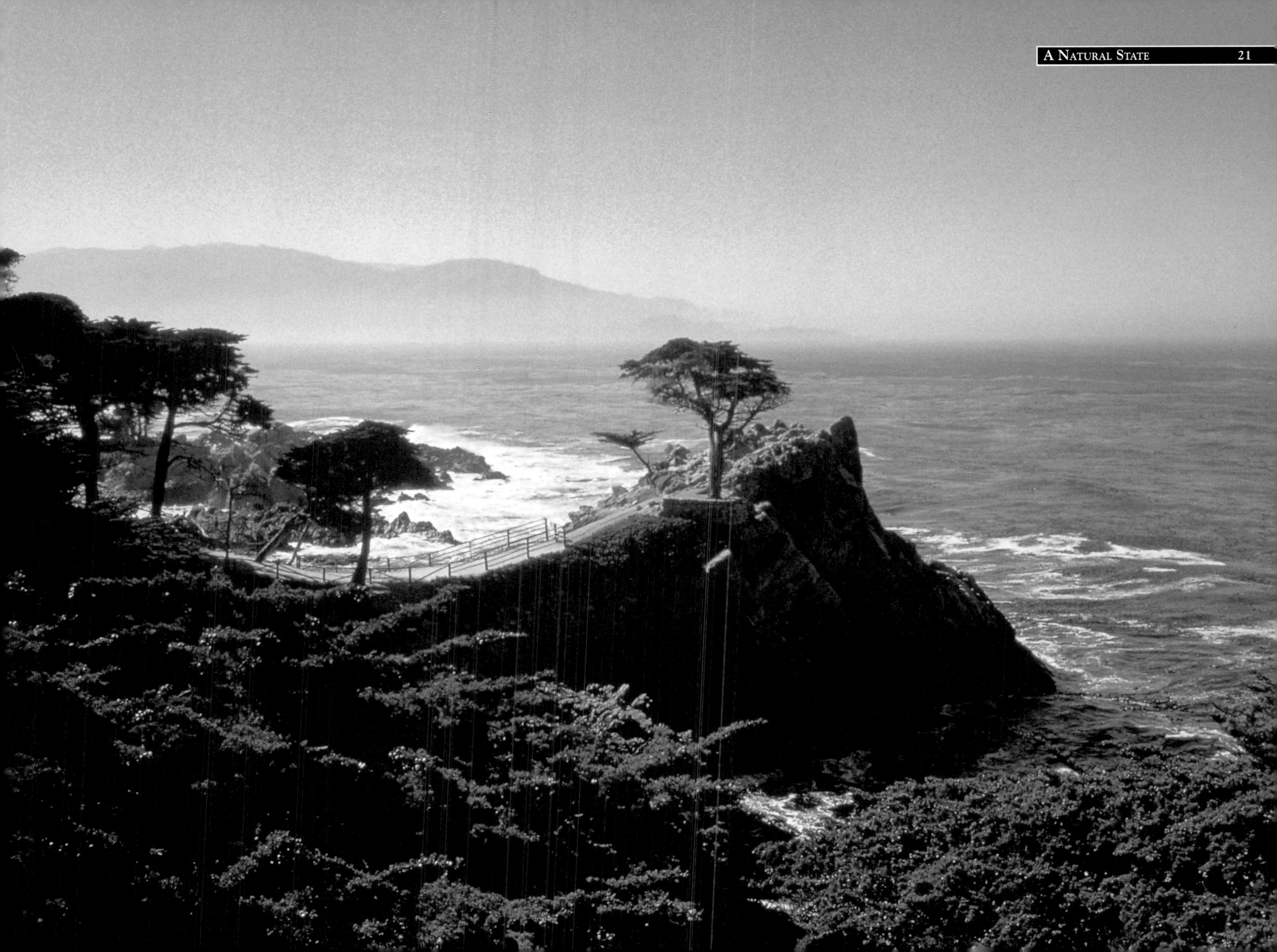

A PLACE

"I had been too long in the calm Sierra pine groves and wanted to hear surf and the cries of seabirds. My son Gen and I took off one February day to visit friends on the north coast. We drove out of the community. A watershed is a marvelous thing to consider: this process of rain falling, streams flowing, and oceans evaporating causes every molecule of water on earth to make the complete trip once every two million years.

Yuba River canyon, and went north from Marysville — entering that soulful winter depth of pearly tule fog running alongside the Feather River and then crossing the Sacramento River at Red Bluff. From Red Bluff north the fog began to shred, and by Redding we had left it behind. ...

This vast area called "California" is large enough to be beyond any one individual's ability (not to mention time) to travel over and to take it all into the imagination and hold it clearly enough in mind to see the whole picture. ... The water cycle includes our springs and wells, our Sierra snowpack, our irrigation canals, our car wash, and the spring salmon run. It's the spring peeper in the pond and the acorn woodpecker chattering in a snag.

IN SPACE

The watershed is beyond the dichotomies of orderly/disorderly, for its forms are free, but somehow inevitable. The life that comes to flourish within it constitutes the first kind of

The surface is carved into watershed's kind of familial branching, a chart of relationship, and a definition of place. The watershed is the first and last nation whose boundaries, though subtly shifting, are unarguable. All public land ownership is ultimately written in sand. The boundaries and the management categories were created by Congress, and Congress can take them away. The only "jurisdiction" that will last in the world of nature is the watershed, and even that changes slightly over time. If public lands come under greater and greater pressure to be opened for exploitation and use in the twenty-first century, it will be the local people, the watershed people, who will prove to be the last and possibly most effective line of defense.

Let us hope it never comes to that."

— Gary Snyder —
A Place in Space

The Rush

It was the shout heard 'round the world –
gold had been discovered in California.
Within months California's population
exploded as the thought of striking it rich
served as a universal magnet.

"Within two years of the discovery of gold in river gravel, gunpowder was blasting the hard-rock fissures. Into the quiet country of the low Sierra – between the elevations of one thousand and four thousand feet – gold seekers spread more rapidly than an explosion of moles."

– John McPhee
Assembling California

The gold lured men from all corners of the world. In the year 1848, some 400 new settlers came to what was then Mexican California. In its new role as a frontier mining town, California drew some 90,000 settlers in 1849, the first major year of the Gold Rush. The 1850 census determined that 24 percent of California's population had come from foreign lands.

It was a difficult journey, whether by sea or land, typically lasting five to six months. Ships on the East Coast faced the task of sailing around Cape Horn to the West Coast, about 18,000 miles. Some gold seekers chose to sail to San Francisco by way of Panama, which required an overland journey through thick jungles, subjecting travelers to

Gold mining was backbreaking work. Picks and shovels lifted and swung as miners loosened the soil. Gold pans and cradles washed the dirt from the ore. It meant long hours standing in an icy stream under a hot sun with only a slight possibility of striking it rich.

From the initial discovery on the American River near Coloma, miners spread to the Feather, Yuba, Bear, Cosumnes, Stanislaus, Tuolumne, Mokelumne, Calaveras and Merced rivers and their tributary streams in search of gold.

Big finds were initially possible. On the Yuba River, the first five prospectors made $75,000 in three months. At Knights Landing on the Stanislaus River, early miners averaged $200 to $300 a day. But for most miners, the typical return was an ounce a day.

Stories of the ease of some of the earliest finds helped fuel the fever. With just the use of spoons, knives, shovels, frying pans, buckets and bowls, men discovered ounce after ounce of gold – most of it just beneath the surface of the land or in stream-

beds. Such methods required little capital investment, although enterprising merchants were quick to make money off goods sold to the miners.

Gold fever struck some, but most struck out. They headed home or turned to commercial enterprises such as tending a bar or running a store – a more certain means of wealth. Food and clothing were in high demand and short supply, producing inflated prices. An onion might cost $1, pickaxes $8 and boots as much as $40. •

s a crew of men deepened and cleared the tailrace for a new sawmill on the South Fork of the American River near Coloma on January 24, 1848, a gleam caught James Marshall's eye – gold had been discovered and life in California would never be the same. Despite the effort by Marshall and the sawmill's owner, John Sutter, to keep it quiet, the news spread. (Sutter himself boasted of the find; he could not keep it to himself.)

By summer, many California communities were abandoned as residents of San Francisco, San Jose, Monterey, Santa Barbara and Los Angeles raced to the Sacramento area foothills to try their luck. Soldiers deserted the Army and crews jumped ship in San Francisco.

California settlements were initially emptied by the rush to the gold fields, but an influx of new settlers from other states and countries, and the money to be made off mining supplies, hotels, saloons and restaurants, reversed this tide.

Sutter, himself an immigrant from Switzerland, arrived in California in 1839. He had established a fort and settlement in Sacramento, which he called New Helvetia (New Switzerland). Over the next eight years, he established a tannery to cure cow hides, a bakery, and a flour mill, among other ventures – not all of which were successful.

In 1847, he contracted with Marshall to build a sawmill on the South Fork of the American River near a place now known as Coloma, an area outside his land grant. Although his sawmill was the source of the Gold Rush, Sutter himself did not strike it rich.

After the discovery of gold at the mill, Sutter turned prospector and took to the hills. But he had little luck. He returned to his fort, but found it in ruins; livestock had been slaughtered for meat; goods and equipment had been stolen; fences had been destroyed. Other ventures failed, and he ultimately petitioned Congress – unsuccessfully – for passage of bills that would provide him with financial aid. Sometimes referred to as the Father of California, Sutter died in 1880 in Pennsylvania. He was 77. •

disease and delays. Many others turned to the overland Santa Fe or Oregon-California trails blazed by the fur traders and other early pioneers. Formidable mountain ranges and harsh deserts lay between these travelers and their destination – the gold fields.

As mining spread from Coloma, hundreds of settlements sprang up: Grass Valley, Auburn, Nevada City, Columbia, Georgetown, Sonora and Placerville, originally known as Hangtown. (A large oak served as the gallows from 1848 to 1857.)

Other camps with colorful names reflect the Wild West atmosphere of those early-day gold fields: Murderer's Bar, Pinchemtite, Rescue, Chili Bar, Fairplay, Hardscrabble, Poverty Hill, Coarsegold, Gold Flat, Gold Springs, and Hardtack.

Other settlements reflected the influx of settlers from other countries: German Bar, French Corral, Dutch Flat, Chinese Camp and Italian Bar. People of every origin – African-

American, Asian, South American, Australian and European – sought California's gold.

Miners were a new force in California. Unlike the Spanish, who had closely followed their traditions many of these pioneers brought new values with them, discarding the conventional way of doing things in favor of new ideas.

Perhaps this was the start of that oh-so-elusive California trait of setting the trends for the rest of the nation. Without question, it was the start of a philosophy of putting California's water to work.

These miners had come equipped with two traits that would begin to shape the state's destiny: technological skills – a burgeoning ability to build succeedingly more elaborate water-works – and a willingness to innovate as they would do anything to get at the precious metal. •

"Gold mining is Nature's great lottery scheme. A man may work in a claim for many months, and be poorer at the end of the time than when he commenced: or he may "take out" thousands in a few hours.

It is a mere matter of chance."

– *Louise Amelia Knapp Smith Clappe* – The Shirley Letters from California Mines

"There is nothing which impresses me more strangely than the fluming operations. The idea of a mighty river being taken up in a wooden trough, turned from the old channel, along which it has foamed for centuries, perhaps, its bed excavated many feet in depth, and itself restored to its old home in the fall, these things strike me as almost a blasphemy against Nature."

– Louise Amelia Knapp Smith Clappe – The Shirley Letters from California Mines

arly gold panning soon gave way to more elaborate means of accessing the precious metal. Some gold seekers built wooden sluice boxes in which they used the power of a stream's flow to separate the gold from the gravel.

Miners built temporary dirt cofferdams, diverting rivers from their natural courses to allow easier access to the riverbed and its gold. In other areas, they moved rivers from their natural channels to the

mining site via wooden flumes or dirt ditches.

These were some of the state's first waterworks. Thousands of trees were cut down and mile after mile of ditches were dug to move the water away from the gold. Miners posted notices on trees to claim water from rivers, establishing the "first in time, first in right" appropriative water rights system that remains in place today with other water rights.

Hydraulic Mining

Beginning in 1850, miners turned from sluice box and long tom to hydraulic mining, the practice of aiming large canvas or iron pipes at hillsides, washing the dirt into a sluice box to expose and recover the ore. The flumes built earlier to divert water away from the gold became the source of power to find the gold. By 1854, more than 4,000 miles of ditches and flumes had been built to deliver water to hydraulic mining operations. Nevada County alone had over 700 miles of such ditches by 1857.

prospecting soon gave way to large-scale company operations.

Hydraulic mining was profitable in large part because the water pressure allowed six men to do the work of 200 miners wielding picks and shovels. But hydraulic mining was destructive. Miners recovered only the gold, letting the remaining dirt and debris known as slickens wash downstream, destroying salmon-spawning beds, raising river bottoms and causing downstream flooding. Some 20 million cubic yards of dirt washed into the American River alone.

Mountains were leveled and canyons were carved by the force of the giant hoses. Hillsides were blasted away, all to uncover the gold. It was perhaps the first industrial use of water in California as individual

"Although in later years the American might lift up his eyes to the hills as he made the valleys bloom, during the Gold Rush he came to the mountains of California as a destroyer. Mining, especially in its more complex phases, devastated the foothills and stripped them of timber. By 1859 much of the Mother Lode had been left a wasteland of caved-in hillsides, heaped debris, and tree stumps. Hydraulic mining, the washing down of mountainsides by high-pressure hoses, poured thousands of tons of mud and silt into the Yuba, Feather, American, Merced, and Sacramento rivers, which ran reddish-yellow with waste."

– Kevin Starr –
Americans and the California Dream, 1850-1915

Yuba River in 1883 following a break in English Dam.

The downstream flooding washed out farms and cities in the valley below and the debris hampered river navigation – leading to a long legal conflict between farmers and miners. Farmers secured a victory in 1884

Located on the remote San Juan Ridge between the south and middle forks of the Yuba River, the North Bloomfield Gravel Mining Co. began its hydraulic mining operations in Malakoff Diggins in 1866. Using high pressure hoses, miners blasted away the hillsides and created a canyon (above

right), discharging a huge volume of mud, sand and gravel into rivers because of the widespread damage it caused, not that hydraulic mining itself was illegal. In fact, some companies resumed hydraulic mining, building log dams to hold back the debris. But these dams failed and by 1900 the industry was dead.

Sawyer's 1884 court ruling was the first in favor of the environment. Today, one can view the remnants of the environmental damage caused by hydraulic mining – a 1,600-acre mine pit – at Malakoff Diggins State Park (left).

with the judicial decision in *Edwards Woodruff v. North Bloomfield Gravel Mining Co., et al.*

Farmers initially turned to the California legislature for help, only to watch the hydraulic mining interests prevail. The North Bloomfield suit was one of several they filed against mines. In his decision, Judge Lorenzo Sawyer of the Ninth Circuit Court in San Francisco ruled that it was illegal to dispose of the tailings into the state's rivers because of the widespread damage it caused, not that hydraulic mining itself was illegal. In fact, some companies resumed hydraulic mining, building log dams to hold back the debris. But these dams failed and by 1900 the industry was dead.

the Yuba River.

Already a flood-prone river, the debris filling the Yuba's bed further increased the risk of flooding in downstream Yuba City and Marysville, and surrounding farm fields.

The Gold Country Today

As one drives the roads today through the Gold Country towns that remain, the vanished landscape of the rush comes to mind only when one spots historical markers along the road. It is hard to imagine the landscape that once existed, to picture the mining camps that have long since vanished. Even where historic towns remain and landmarks have been restored, the true air of the time – the roughness, the hangings, the liquor, the fights, the hardscrabble existence – is difficult to bring to mind.

Instead there is a genteel look of bed-and-breakfasts, art galleries, antique shops and restaurants set against the golden hills – gold today with dried, wheat-colored grass instead of the ore that gave the region its singular place in California history. Small museums and a replica of Sutter's mill, state parks of preserved towns such as Bodie, the still-standing flumes and former mines such as the Malakoff Diggins hint at the past. So,

too, do the mounds of smooth river rocks that lie on either side of many of the waterways that are now the destination for those seeking excitement of a different type – white-water rafting.

Still, tourists can test their skills with a gold pan at Coloma – the place where it all began. This reenactment offers perhaps the clearest taste of the excitement of yesteryear – the potential of instant riches. It is the thrill of entertaining the possible "what if" that most distinctly replicates the gold fever of the '49ers. •

Arteries of Commerce

"The other side was the reeling, rolling, yellow Sacramento River – a forbidden menace and a fascinating vision. That's where the steamboats plied, the great, big, flat-bottomed cargo and passenger boats, some with side wheels, some with one great stern wheel. I did not know, I did not care, where they went. It was enough that they floated by day and whistled by night safely on that dangerous muddy flood which, if it ever got a toy in its grip, would roll him under, drown him, and then let his body come up all white and still and small, miles and miles away."

– Lincoln Steffens –
Boy on Horseback

rior to the railroad, the freeway and the airplane, rivers served as the nation's arteries of commerce. While early federal policies encouraged the use of California's rivers for commerce and laws cleared the way to make navigation a top priority, many waterways proved to be unnavigable because of unpredictable flows.

Two exceptions were the Sacramento and San Joaquin rivers, which provided vital passageways linking the ocean port cities of San Francisco and Oakland with the Gold Country. When ship traffic was at its height, riverboats and barges cruised upstream on the Sacramento River as far as Red Bluff, 300 miles north of Sacramento.

On the San Joaquin, ships once traversed the river between Stockton and Fresno twice a week. Even the Feather River had waterfront developments for commerce in Oroville and Marysville. ●

"We kept on our slow and winding way, often on bars and shoals that took long to get over. A wide plain borders the river on each side. We caught distant views of the mountains, but generally we saw only the river and its banks, which were more or less covered with trees – willows, cottonwoods, oaks, and sycamores – with wild grapevines trailing from them."

– William H. Brewer
Up and Down California in 1860-1864

New York to San Francisco. The boat was said to have made $60,000 a month in her first year of operation.

Smaller steamships transported freight and people up the Sacramento and Feather rivers to Red Bluff and Marysville. As early as 1851 so many ships arrived at the Marysville docks there wasn't enough room to discharge cargo. Four years later, some 200 tons of freight arrived daily via steamship and barge.

High demand and high rates led to other vessels since, at the time, California had to import virtually all its food and merchandise. By 1850, as many as fifty steamers were making the trip between Sacramento and San Francisco. One of the most profitable and luxurious of these boats was the 4,500-ton *Senator*. In 1849, San Francisco to Sacramento freight charges on the *Senator*, called a floating palace, were $40 per ton — more than the cost of shipping from

here motorboats ply the waters today, commercial steamships were once the primary link between the Sacramento Valley and the rest of the world. Steam navigation on the Sacramento River began in 1849 when several flat-bottomed steam scows and side-wheelers were put into commission. Among the first ships were the *Washington*, *Edward Everett*, *Jr.*, and *Sacramento*, ships which, according to early records, were rather unreliable.

Compared to the sailing vessels they replaced, however, they were a much faster way to ferry goods and people from San Francisco to Sacramento. Where sailing ships would take three to ten days to make the trip, the larger steamboats eventually established a ten to twelve hour schedule of service. Fierce competition led to steamboat races to see who could be the fastest. Sometimes overstressed boilers exploded, with disastrous consequences.

Debris from hydraulic mining and the development of the railroad slowed riverboat traffic until the 1900s when the federal River Navigation Project was initiated. For the Sacramento Valley, the resurgence of the steamboats led to a record number of passengers and an enormous amount of freight transported in 1925 – 1,366,780 tons. •

Romance of the Riverboat

Mark Twain's story of Huckleberry Finn and his trip down the Mississippi River illustrates one of the main appeals of the riverboat – the lure of adventure. The river was the avenue to somewhere else; the boat provided the chance to see what was around that bend. Drawing on his experience as a river pilot, Twain had Finn travel down the Mississippi in search of a new life. For some – just as for Twain's Finn – the river and the steamer offered passage to a new life.

For others, the riverboats were the most practical and luxurious way to travel from California's small inland towns to the city of San Francisco for a welcome excursion. Live music was provided by bands. The 1,000-passenger *Chrysopolis* featured elegant cabins and murals. The *Senator* had rosewood staterooms and four bridal suites. The *Fort Sutter* and the *Capital City* advertised that their staterooms included baths.

Perhaps the most famous of these San Francisco-Sacramento steamers were the palatial *Delta King* and *Delta Queen*. Well into the twentieth century the overnight cruise on these boats remained a popular mode of transportation. Sacramento Valley residents would take the *Delta Queen* to San Francisco, known as "the city," spend the day in town, and return at night on the *Delta King*. These deluxe paddle wheel steamers were opulently furnished with elaborately turned moldings, red plush upholstery, marble-topped tables and gilt-frame. "Riverboats were the links that tied Delta residents to the outside world.

The boats' calls were social events. Whole towns rushed to the wharves to see who was a'comin' visiting, while Chinese, Japanese and turbaned Sikhs trooped ashore, almost unnoticed. Here, one picked up on news, gossip and rumors – and snooped to see who was off to the City on a spree," Richard Dillon wrote in *Delta Country*.

The *Delta King* and *Delta Queen* were the pride of Sacramento and remained in use until World War II, when they were put into wartime service. Later, the *Delta Queen* became a cruise ship on the Ohio and Mississippi rivers while the *Delta King* fell into disrepair. Today, the restored *Delta King* serves as a popular restaurant/hotel moored on the Sacramento River at Old Sacramento. Here it retains a romantic tie to a slower paced life and a reminder of California's first arteries of commerce – its inland waterways. •

An Artery to the World

Much has changed since the first ships docked at Sacramento's wharf some 150 years ago. The rush to the gold fields forever transformed a frontier outpost into a bustling port city. A century and a half later, the Sacramento and San Joaquin rivers remain vital links to ocean-bound ships and the cargo they carry.

By way of special deep-water channels dredged by the U.S. Army Corps of Engineers, giant ships regularly dock at ports in both Sacramento and the city of Stockton.

Stockton's deep water port opened in 1933. Ships up to 900 feet long and weighing as much as 80,000 tons (depending on load size) can navigate the channel that stretches 75 nautical miles to San Francisco Bay. Exports include such agricultural products as grain, wheat seed and rice while imports include cement, machinery, steel and sugar.

On June 29, 1963, 5,000 spectators turned out for the grand opening celebration of the new port of Sacramento. That day, the motor vessel *Taipei Victory* became the first ocean going ship to dock in Sacramento since 1934. Civic leaders had campaigned for their own deep water channel and port for some thirty years. Today, 1.3 million tons of goods are shipped in and out of the port annually on the 79-mile channel linking the Golden Gate with this interior city. ●

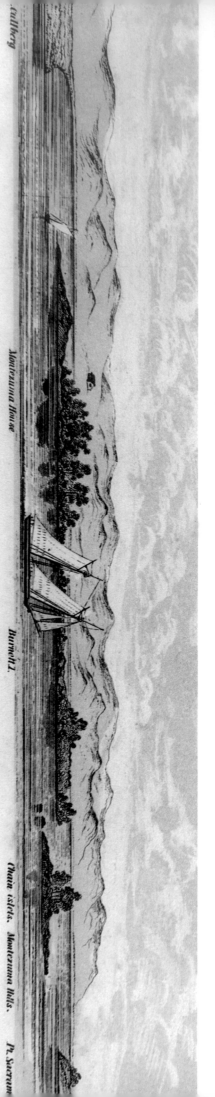

Entrance to the Sacramento River

Culberg   Montezuma House   Burnett.   Chain islets.   Montezuma Hills.   Pt. Sacram[...]

The Delta

Yesterday's Delta bears scant resemblance to today's maze of waterways surrounding fifty-seven reclaimed islands. Then, the Delta was comprised mostly of intertidal wetlands with few, if any, elevated landforms. Steamships traveling to and from San Francisco and Sacramento regularly followed a deep channel through the Delta still known today as Steamboat Slough.

Here and there deposits of sediment formed small, natural levees. Early travelers noted the thick clouds of mosquitoes, the abundant beavers and their islets and the tall and broad expanses of tules, along with a riparian forest of sycamores, cottonwood, alder and willow.

Along Delta waterways much of the riparian forest was harvested to provide fuel for the wood-burning steamers as well as firewood and building material for Sacramento, Stockton and San Francisco. In fact, one of the earliest Delta industries was wood cutting. U.S. Navy Commodore Cadwalader Ringgold, who charted the Sacramento River in 1849-1850, described it thus: "A lively scene is presented to persons passing up and down the river; at almost every bend and turn, the woodcutter is seen and the pleasant sound of his axe heard, with hundreds of cords of wood convenient for transportation."

At Steamboat Slough, boats would back up to the banks to "wood up." Woodcutting for steamships continued well into the 1870s. Land cleared of the native cottonwood and sycamores, was then transformed into orchards and other crops. Early diaries show that by the late 1800s and early 1900s, much of the riparian growth along Delta sloughs and the rivers far upstream had been removed.

Farming in the Delta began in the 1850s, with settlers encouraged by early federal laws to transform the marshland into farmland. Between 1852 and 1857, portions of Grand, Rough and Ready, Andrus, Roberts and Union islands were reclaimed through construction of levees. Farmers built small dikes around areas of the marsh, draining the land of water and creating an island. Early levees were only 3 feet tall, composed of the area's natural peat soil – washing away in high water. Each flood would prompt a rebuilding, as farmers built higher and higher levees.

Additional legislation encouraging the reclamation of wetlands and

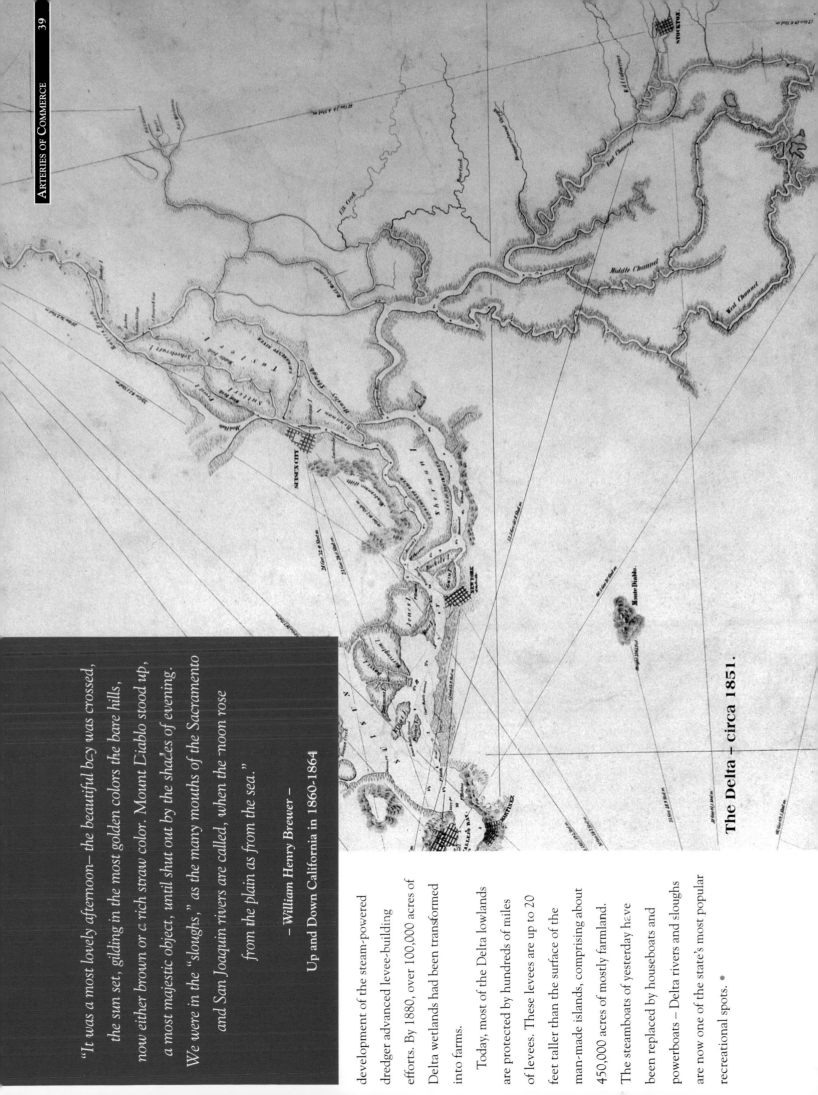

The Delta – circa 1851.

"It was a most lovely afternoon— the beautiful bay was crossed, the sun set, gilding in the most golden colors the bare hills, now either brown or a rich straw color. Mount Diablo stood up, a most majestic object, until shut out by the shades of evening. We were in the "sloughs," as the many mouths of the Sacramento and San Joaquin rivers are called, when the moon rose from the plain as from the sea."

— *William Henry Brewer* —
Up and Down California in 1860-1864

development of the steam-powered dredger advanced levee-building efforts. By 1880, over 100,000 acres of Delta wetlands had been transformed into farms.

Today, most of the Delta lowlands are protected by hundreds of miles of levees. These levees are up to 20 feet taller than the surface of the man-made islands, comprising about 450,000 acres of mostly farmland. The steamboats of yesterday have been replaced by houseboats and powerboats – Delta rivers and sloughs are now one of the state's most popular recreational spots. ●

The Inland sea

In their natural state, California's Sacramento River and its tributaries regularly overflowed their banks. Water levels rose with winter storms or spring snowmelt, and a huge seasonal wetland or "inland sea" would form, stretching from Mount Shasta to the Delta.

Only in the late spring or early summer would the waters drain from the land, leaving behind a deposit of rich river silt. Layer upon layer of this sediment formed a fertile bed of soil.

Miners who failed to find gold became farmers, transforming the valley floor into a sea of crops. But the seasonal wetland that was the valley made farming difficult.

The Native Americans had lived synchronistically with the valley's recurring sea, moving between lower and higher ground as nature dictated. Such a nomadic lifestyle was in direct contrast to the driving force to settle, to colonize California. For these early settlers, the inland sea became something to fight – a force they must somehow hold back to protect their livelihoods.

Rather than retreat and move to higher ground, residents of towns such as Sacramento, Marysville and Yuba City chose to resist the rivers. ●

sea of floodwaters spread far and wide in the Sacramento and San Joaquin valleys during the flood of 1861-1862. For decades, it stood as the worst-recorded flood in the valleys' history as unprecedented rainfall caused the rivers to overflow their banks.

For the city of Sacramento, the flood of 1861-1862 was known as the Great Calamity, with two floods a month apart. The first occurred December 9,

1861 when a levee on the American River broke, sweeping a wave of water into town and burying city streets under twenty feet of water. The floodwaters became trapped by the remaining levee, forcing residents to cut a hole in it to drain water into the Sacramento River.

On January 22, 1862, a levee along China Slough collapsed, inundating the city once again. The state legislature fled to San Francisco and the floodwaters lingered until March.

"Such a desolate scene I hope to

never see again," William Henry Brewer wrote on March 9, 1862. "Most of the city is still under water, and has been for three months. ... the new Capitol is far out in the water – the Governor's house stands as in a lake. ... I left the city and as I came down the river, saw the wide plain still overflowed, over farms and ranches – houses here and there in the waste of waters or perched on some little knoll now an island." ●

"Nearly every house and farm over this immense region is gone. There was such a body of water – 250 to 300 miles long and 20 to 60 miles wide – that the winds made high waves which beat the farm homes in pieces. America has never before seen such desolation by flood. ... But the spirits of the people are rising, and it will make them more careful in the future."

– William H. Brewer *Up and Down California in 1860-1864*

"Soon after the Gold Rush … thousands of the people who came to Central California followed a brief fling at the mines by moving down from the mountains to settle in the fertile Sacramento Valley. Here they shortly encountered a gravely threatening natural phenomenon. They discovered that during the annual winter cycle of torrential storms that for millennia have swept in from the Pacific, or in the season of the spring snow melt in the northern Sierra Nevada, the Sacramento River and its tributaries rose like a vast taking in of breath to flow out over their banks onto the wide Valley floor, there to produce terrifying floods. On that remarkably level expanse the spreading waters then stilled and ponded to form an immense, quiet inland sea a hundred miles long, with its dense flocks of birds rising

Tulare Lake

Snowfall in the Sierra fed the seasonal body of water known as Tulare Lake. In most wet years, the shallow lake covered some 200,000 acres at its peak, shrinking quickly as winter runoff from the Kings, Kaweah, White and Tule rivers slowed and summer temperatures rose.

Well into the late nineteenth century, Tulare Lake was the largest body of fresh water west of the Mississippi River. Thick stands of tules — bullrushes 8 to 15 feet tall — lined the lake's shore. Early explorer William Henry Brewer forced his way through

to view the lake. Malaria posed a problem in the summer as the standing water attracted scores of mosquitoes. Early on, Native Americans built rafts from the tules. In later years steamboats and commercial vessels seeking waterfowl, fish, frog legs and turtles plied the fluctuating waters. They also moved cargo between Corcoran and Kettleman City.

The lake reached its maximum recorded size during the floods of 1861-1862, engulfing nearby Kern and Buena Vista lakes and swelling to nearly 500,000 acres.

Flood control dams built in the mountains on the lake's four tributary rivers have all but ended the formation of this seasonal lake. The lake bed itself now consists of row after row of crops such as cotton and safflower.

Since the twentieth century, the lake has occasionally reappeared in years of high runoff. In 1969, flooding again made Tulare Lake one of the state's largest lakes. Flood flows brought back the lake in 1978, 1983 and, most recently, 1997 — albeit at a much smaller size than in the past. •

abruptly to wheel in the sky and its still
masses of tule rushes stretching from
the Delta to the Sutter Buttes and
beyond. Not until the late spring and
summer months would it drain away
downstream. For the better part of the
next several generations, embattled
farmers and townspeople struggled to
get control of their great river system
so they might live in safety on the Valley
floor and put its rich soils to use. In our
time, after that long labor, we observe in
the Sacramento Valley a literally remade
environment, a creation of artifice, a
produced object shaped into disciplined
and rational form after many fumbles
and misdirections, and decades of
humbling trials and errors."

– Robert Kelley –
Battling the Inland Sea

"From Tulare Lake come the turtles that make the
rich turtle soups and stews of San Francisco hotels and restaurants.
It is the western pond turtle common in the fresh water ponds.
The Italians call it Ella-chick. These turtles are sent in sacks to
San Francisco. During the season more than 180 dozen found
a ready sale at the bay."

– History of Kern County, 1883 –

Tulare Lake during the 1997 floods.

From Swampland to Farmland

The seasonal wetlands of the Sacramento and San Joaquin valleys, Tulare Lake region and the Delta were often referred to as tule lands. In the Sacramento Valley, these thick, high-stemmed rushes ran along both sides of the Sacramento River down much of the valley into the Delta. Before this land could be farmed, the tule had to be broken down, burned and then plowed.

Initially, these lands could be farmed only seasonally because of the unpredictable Sacramento and San Joaquin rivers. First to be cultivated were the so-called rim lands at the wetlands' edge. Planting with the cycle of the river, those known as rim farmers found the soil to be of good quality, for it was formed by silt deposited during frequent floods. The rim lands were among the first to be protected by levees — small, low, crude dikes.

The transformation of the inland sea to the irrigated fields one sees today began shortly after California became part of the United States. The "swamp and overflowed lands"

were identified on maps of the time were initially in federal ownership. In 1850, Congress passed the Arkansas Act, transferring title to these lands to the states. The California legislature, in turn, authorized the sale of these lands for as little as $1 an acre to private landowners, who were obligated to drain and reclaim the tracts.

These lands were among the least expensive, offering frustrated gold miners an affordable way to become farmers. And while agricultural production offered much of the impetus to drain these wetlands, public health was a major concern — these seasonal wetlands were prime breeding grounds for mosquitoes, carriers of the dreaded malaria. Political and technological changes accelerated reclamation. Laws passed in the 1860s encouraged landowners to form self-taxing reclamation districts to finance levee construction and drainage canals.

In some instances, reclamation was conducted on a large-acreage basis by private companies, which later sold the former swampland to farmers at a

handsome profit. Reclamation in the Sacramento-San Joaquin Delta proceeded at a faster rate than elsewhere in the Central Valley because of private capital and a well-organized effort to dike and drain the islands. By 1880, over 100,000 acres of Delta had been reclaimed. By 1900, some 235,000 acres had been reclaimed. The 1907 and 1909 floods inundated nearly all of the reclaimed Delta islands.

Development of the steam-powered dredge — especially the clam shell dredge — allowed for faster construction of larger levees than those built by the mostly Chinese workforce. The dredge dug down into the riverbed, extracting mud, then pivoted over to the side of the river to deposit it.

Not everyone was against swamps. According to Robert Kelley, author of Battling the Inland Sea, there were some who questioned the Swampland Commission's dream of draining and drying out the seasonal wetlands. Farmers, for one, recognized the role seasonal floodwaters played in providing fertile soil. ●

Value of Wetlands

Swampland reclamation activities, flood control projects and the growth of cities reduced the amount of habitat for waterfowl, which, in turn, decreased the number of birds. But reclamation alone was not the only problem. Early hunting and egg-gathering practices also decimated bird populations. From 1850-1856 alone, 3 to 4 million eggs were taken from seabirds nesting on the Farallon Islands. Market hunters went into the tules to shoot wild ducks and geese, and hundreds of thousands were sold to the restaurants and hotels in San Francisco.

Eventually, national interest in migratory waterfowl prevailed and the first federal wetlands protection programs were initiated. Land was set aside in the Central Valley and Klamath River Basin and managed to support seasonal and year-round bird populations. Scientists who gained new appreciation for their value created a new term in the 1950s:

"wetland" to replace the more derogatory "swamp."

Despite increasing awareness of the importance of wetlands, conversion continued in the fast-growth, post-World War II decades of the '50s, '60s, and '70s. It wasn't until the 1980s and 1990s that federal and state officials began to encourage the preservation of existing wetlands through such policies as "no net loss."

Of the approximate 450,000 acres of wetlands that remain, most are located in five regions: the Central Valley, Humboldt Bay, San Francisco Bay, Suisun Marsh and the Klamath Basin.

In general they are divided into two categories, the saltwater or brackish wetlands of the coastal regions and the freshwater wetlands of the Central Valley and Klamath Basin. These wetlands provide vital habitat for birds and animals, including the ducks and geese of the Pacific Flyway. •

Flood Fights

Sandbags... Even with one of the most sophisticated flood control works in the world, crews fighting the 1997 Pineapple Express floods turned to more than 100,000 sandbags to save one Sacramento Valley town from disaster.

It was January 6, 1997, the middle of the worst flooding in four decades. After eleven days of torrential downpours throughout most of California, a levee along the Sutter Bypass broke. Officials estimated that the floodwaters would reach the small community of Meridian — including a hundred homes, a church, a school, a post office and a café — in twelve hours. As the 400 residents evacuated their homes, emergency crews began building a temporary levee (left).

Engines roaring, bulldozers and backhoes moved 50,000 cubic yards of dirt into place, forming a 4,700-foot berm around three sides of the town. Sheets of plastic were laid over the dirt to protect against erosion and crews then began to layer it with thousands and thousands of sandbags – one at a time.

Officials cut a new hole into the Sutter Bypass levee to drain the flood waters back into the bypass and away from the town. Then, they waited.

The water level rose, lapping at the sides of the makeshift levee, whose height ranged from 5 to 8 feet. The winds blew, sparking fears that the berm would give way to erosion, but it didn't.

Several days later, the temporary levee was still standing, the town was still dry and the water level began to drop. The New Year's Flood of 1997 was subsiding, but the cleanup work had just begun.

While Meridian was saved, many other communities were not. Fed by rain and melting snow, rivers tore holes through levees, spilling muddy, debris-ridden waters across farm fields, roads, driveways and into living rooms. Statewide, some 23,000 houses were damaged, 300 square miles of farmland were swamped and nine people lost their lives in the 1997 flood. ●

wollen rivers followed two days of rain in the Sierra Nevada and on January 9, 1850 the Sacramento River swamped the young city of Sacramento. With no levee to protect it, the waterfront town soon found the river running through its streets about a mile back from the river's edge. Boats entered the city hotel by way of its second-floor windows.

Early Sacramento pioneer John Sutter had built his fort on high ground away from both the Sacramento and the American rivers, but the '49ers' rush to settle the city caused it to grow out from the fort toward the rivers, which afforded easy transportation. After the flood, four-fifths of the city was under water.

As miner Jacob Stillman wrote in a letter to his wife, "All sorts of means are in use to get about – baker's troughs, rafts and India-rubber beds. There is no sound of gongs or dinner bells in the city. The yelling for help by some men on a roof or clinging to some wreck, the howling of a dog abandoned by his master, the boisterous revelry of men in boats who find all they want to drink floating free about them, make the scene one to never be forgotten. After dark we see only one or two lights in the second city of California."

It was the first of many floods to engulf the city, for its location at the confluence of two major rivers – the American and the Sacramento – tied it inextricably to such a fate.

Following the flood, Hardin Bigelow rallied the city to build a levee. Only 3 feet tall and 12 feet wide at its base, the levee was viewed in triumph – and Bigelow elected mayor.

Victory was short-lived, however, as the levee failed in the floods of 1852. ●

With only dirt levees to hold back the forces of the rivers and the tide, the Sacramento-San Joaquin Delta is prone to flooding. Since 1967, some twenty-nine levee breaks have occurred. For some of these man-made islands, such flooding is a regular occurrence.

In June 1972, the San Joaquin River cut a hole 75 feet deep and 500 feet wide through the levee protecting Andrus and Brannan islands. Some 13,000 acres were flooded, including the city of Isleton.

Delta islands are much like holes, the surface of the land is well below the top of the levee. When a levee breaks, the water rushes in. In the case of this break, it was twenty-four days before it was closed and several more weeks before portable pumps succeeded in pumping the island dry. •

The state's major flood works — in the Central Valley, along the north coast and in southern California — have saved untold lives and billions of dollars in property since they were built.

Floodwaters can be captured and stored for later release.

Floodwaters can be directed away from areas of concentrated population to more rural land. In short, floodwaters can be managed, reducing the loss of life and property damage. No system, however, is 100 percent effective at controlling floods.

Consider 1997.

In the worst flood event since 1955, a series of warm tropical storms hit California in late 1996 and early 1997. Known as the Pineapple Express, the first of these back-to-back warm storms arrived December 26. By the time the rains let up on January 6, the combination of more rain and less snow (because of warmer weather) and a melting snowpack caused rivers throughout the northern and central part of the state to overflow their banks as reservoirs made record releases.

Among others, the Feather, San Joaquin, Bear, Yuba and Cosumnes rivers broke through their levees, flooding the Central Valley.

In southern California, the rains caused massive mud slides and flash floods. By the end of January, forty-eight of the state's fifty-eight counties had been declared disaster areas with an estimated $2 billion in damages caused by the New Year's storms. •

> *"Floods are acts of God: Flood damages result from acts of men."*
>
> *— Charlie Casey — Friends of the River*

"Those early levees weren't worth a lick;
first wet spring and they gave 'way right and left.
The wind could get that water churned up,
get it lapping at the dirt banks and chewing them away
faster than a man could fill them in. Farmers tried everything
from sandbags to wrecked automobiles to hold the water back
and plenty of times they were flooded out anyway."

– Nick Zachreson –
The San Joaquin Valley

Living in a Floodplain

"Land use planning" was not a stock phrase when California was first settled. Rivers served as the arteries of commerce, so it seemed only natural to settle by their banks. The fact that these low-lying areas were historic floodplains was no deterrence; the effort to control nature was the overriding sentiment.

With dams and other flood control features in place, fears were placated and growth continued. Small towns became cities, and cities became metropolitan regions. The modern-day inhabitants of urban California lost touch with the natural cycle of floods, made worse by the paving over of land that once absorbed water, and by upstream development that increased downstream flows.

The philosophy was to bridle the rivers to maintain these areas. Government funding financed stronger levees, bigger dams and more sophisticated flood control systems.

Only as the societal and financial costs of recent floods reached epic levels did attitudes begin to shift. While many who were flooded out in 1995, 1997 and 1998 chose to rebuild, others began to take advantage of new government programs designed to buy out residents and return flood-prone areas to their natural state. There also is interest in modifying some levees to allow rivers to create meander belts. •

"As population has increased, men have not only failed to devise means for suppressing or escaping this evil, but have, with singular short-sightedness, rushed into its chosen paths."

— William McGee —
The Flood Plains of Rivers, 1891

The 1955 Flood

Christmas 1955 brought the biggest flood to the Sacramento Valley in ninety years. Thousands of people were evacuated as a series of warm storms brought unprecedented rain to the region. Displaced residents took shelter in emergency refugee centers, where the holiday spirit prompted donations of clothes and toys.

Hardest hit was Yuba City, where a Feather River levee broke, inundating houses up to their rooftops and flooding thousands of acres of farmland. Statewide, sixty-four people were killed, and the flood led to the

campaign for the State Water Project.

In Sacramento, streets were swamped because the saturated ground could not hold any more water. Although some local creeks flooded outlying areas, the historical menace of the American River was constrained by the newly completed Folsom Dam. At the dam's May 1956 dedication ceremony Gov. Goodwin Knight noted the "sorrow that this man-made structure has already saved us." ●

"It is easy to forget that the only natural force over which we have any control out here is water, and that only recently. In my memory California summers were characterized by the coughing in the pipes that meant the well was dry, and California winters by all-night watches on rivers about to crest, by sandbagging, by dynamite on the levees and flooding on the first floor."

– Joan Didion –
The White Album

Oroville Dam under construction during high flows in January 1966.

By 1852, there were 25,000 Chinese in California. Most of them were in the mines, although they were not allowed equal access in staking their own claims. Early laws were designed to discourage the Chinese and other foreigners from mining.

Despite racial discrimination, the Chinese helped build California. Their labor was vital in construction of the Southern Pacific Railroad, linking the state with the rest of the country.

Chinese laborers also helped construct levees in the Delta to reclaim this marshland and turn it into productive farmland.

The Delta town of Locke became the nation's only Chinese-built, Chinese-inhabitated town. Lee Bing leased land from the estate of George Locke. In its prime, Locke had a hotel, bakery, candy store, barbershop, mill, gambling dens, and a Chinese language school. ●

Levees were the first line of defense built against the flood waters of the Sacramento and San Joaquin valleys. The force of the rivers overrode these small levees with ease; residents responded by building taller levees.

In some cases, individual land-owners wound up fighting one another instead of the river as each tried to build a higher levee, protecting one parcel at the expense of another.

When they joined together to create an integrated system rather than building individual levees of varying heights and quality, they began to succeed. With stronger levees in place, valley farmers could plant permanent orchard crops rather than annual wheat and other field crops, increasing the value of the farmland and its harvests.

With the second line of defense in place — upstream dams and reservoirs and a bypass system to divert water out of the rivers in times of high flows — the valley prospered. No levee is guaranteed, however. ●

The wide floodplains of the natural river system in the Central Valley have been replaced with mile upon mile of levees. Levees have been built since the turn of the century as settlers sought to reclaim the land for farming. Today, some 200 miles of rip-rapped levees line the Sacramento River.

Levee construction is only the first step. Levee maintenance is vital to ensure that these earthen dams hold during times of high flow when rivers swell. The wakes from passing boats can cause levees to erode, weakening the barriers that protect life and property from floods. To reduce erosion, officials often use rip-rap, large rocks placed along the water's edge. •

he Salton Sea was formed by the joint forces of man and nature some ninety years ago when the Colorado River breached the levee of an early irrigation diversion channel. Cutting a new channel through the Imperial Valley, the river poured its full fury into an ancient seabed known as the Salton Sink.

The 1905 flood continued for sixteen months as the small breach widened to a mile. The floodwaters inundated a company formed to mine the salt from the bottom of the sink and ship it east in Southern Pacific

Railroad cars. By July 1906, these workers' homes, the machinery, the railroad track and the station were all under 55 feet of water.

As Edgar L. Larkin wrote in the October 1906 issue of Cosmopolitan magazine, "Soon after passing Mecca, someone exclaimed, 'Oh, look! Water in the desert!' Sure enough! There was the sea. A brisk wind was kicking up whitecaps, and these dashed against clumps of sagebrush. A strange combination – sagebrush and water!"

After several attempts to close the gap, an around-the-clock, fifty-four-

day effort in which 6,000 carloads of rock and gravel were poured into the breach succeeded in turning the Colorado River back toward the Gulf of Mexico.

Today's Salton Sea occupies the bed of ancient Lake Cahuilla. Estimated at twice the size of the sea, this lake took some twenty years to fill and sixty years to evaporate. The legacy of this lake can be seen in the Imperial Valley today: the sand dunes, the seashells, and the bathtub ring high-water mark on the Santa Rosa mountains west of the Salton Sea. •

"The declining sun suddenly emerged from behind a peak, its slanting rays illuminating the whole interior.

I saw it all for five minutes – every outline of the ancient shore and of the modern sea."

– Edgar L. Larkin –
"Wonderful New Inland Sea"
Cosmopolitan magazine

At the end of 1915, a series of years with below-normal rainfall had shrunk San Diego's water supply reservoirs. Drought was never very far from the minds of residents in the arid southern California region, and there was concern that the new year could bring even drier conditions.

When a rainmaker by the name of Charles Hatfield offered to produce 40 inches of rain at the city's Morena Reservoir, officials figured they had nothing to lose – especially since the first 30 inches would be free. At the reservoir some 60 miles away from the city near Laguna Mountain, Hatfield built a tower. At the top of the tower he placed a mixture of secret chemicals said to have produced rain for cities throughout the West.

On January 14, 1916, the rains began. By January 16, they had become a torrential downpour and the San Diego River rose rapidly. At Morena Reservoir, 12 inches of rain fell in four days. High flows in the region's creeks and rivers washed out bridges, and flooded farmland and homesteads. By January 20, the Sweetwater Reservoir was nearly full.

The rains eased for a day or two, but began again on January 24. The bridge across the San Diego River at Old Town was swept away. Six inches of rain had fallen at Sweetwater Reservoir by January 27, and water reached the top of the dam and flowed over the parapet. As a wall of water 3 1/2 feet high cascaded onto the valley floor, a 90-foot rock and earth abutment of the wing dam was washed away.

Main Sweetwater Dam remained in place, but the Lower Otay Dam was not as fortunate. Here, the water had risen more than 27 feet in ten days and on January 27, the dam burst, sending an unchecked wave downstream. Only the sides of the dam were left after the flood. Water swept over the fields, livestock and homesteads of the Otay Valley, killing at least fourteen people. Also hard hit was the Mission Valley area (above left), victim of subsequent flooding in 1980 (below left).

Some eighteen people were killed throughout San Diego County in the floods of 1916, which the *San Diego Union* declared the worst such storm in thirty-two years. •

Improving Predictions

A combination of late snow storms followed by triple-digit temperatures generated unprecedented spring runoff in the Colorado River in 1983, and water flowed down the spillway of Hoover Dam on July 25 for only the second time since its completion in 1941.

Such high flows were an all too familiar experience for early settlers of the basin, for this was a river whose historic force carved canyons with ease. Some referred to the 1983 flood as a man-made disaster because the federal officials who operate the dams kept releases below peak flows and did not make enough room in the reservoirs. Yet in a year when snow fell as late as May and 100-degree days followed, others said it was only a question of when the river would flood – not if.

The spill continued until September 6, and the river's high flows washed

"'Earthquake?' [Ronald] Reagan said, shaking his head. 'That's what the public worries about. ... Floods are the thing we worry about most – they're the scariest. There's not a heck of a lot you can do about flash floods, and there's no warning. The ones we get push around boulders the size of vans. And a lot of people are living in certain flood paths.'"

– *Richard Reeves*
"Vulnerable"

out beaches and rapids in the Grand Canyon and inundated recreational campsites, farmland and homes in Arizona, California and Mexico.

Releases from Hoover Dam continued well into 1984 and the incident prompted several changes in project operations and forecasting. Computer modeling, aerial surveys and calibration of remote snow stations have improved runoff predictions. Such information is used to make earlier releases of water, as warranted, to create flood space in the reservoir.

The only other time Hoover's spillway has been used was during tests after the dam was completed. •

"In the winter of wet years the streams ran full-freshet, and they
swelled the river until it raged and boiled, bank full, and then it was
a destroyer. The river tore the edges of the farm lands and washed
whole acres down; it toppled barns and houses into itself, to go floating
and bobbing away. It trapped cows and pigs and sheep and drowned
them in its muddy brown water and carried them to the sea.
But there were dry years too, and they put a terror on the valley.
The water came in a thirty-year cycle. There would be five or six
wet and wonderful years when there might be nineteen to twenty-five
inches of rain, and the land would shout with grass. And then the dry
years would come, and sometimes there would be only seven or eight
inches of rain. The land dried up and the grasses headed out miserably
a few inches high and great bare scabby places appeared in the valley.
The land cracked and the springs dried up and the cattle listlessly
nibbled dry twigs. Then the farmers and the ranchers would
be filled with disgust for the valley. People would have to haul
water in barrels to their farms just for drinking.
Some families would sell out for nearly nothing and move away.
And it never failed that during the dry years the people forgot about
the rich years, and during the wet years they lost all memory of
the dry years. It was always that way."

— **John Steinbeck** —
East of Eden

Sutter Bypass during the 1997 floods.

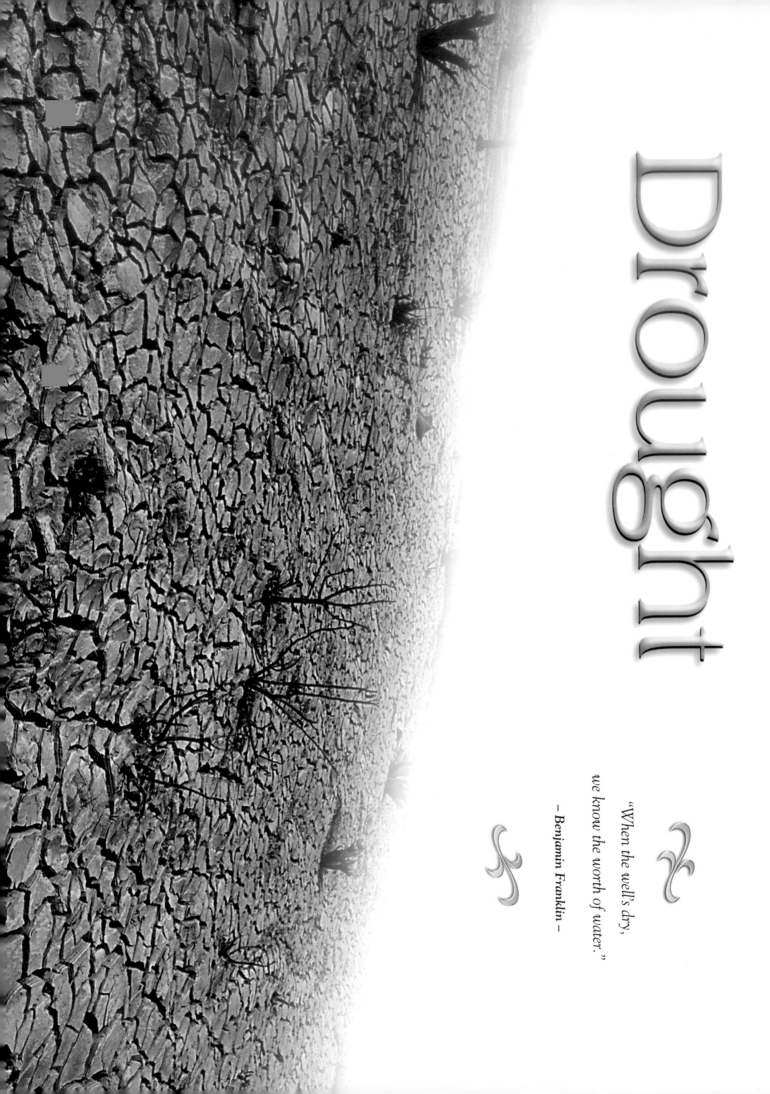

Drought

"When the well's dry,
we know the worth of water."

— Benjamin Franklin —

The Dust Bowl epitomizes drought. A vivid image comes to mind when one thinks of this 1930s drought in which some 100 million acres in the South Great Plains Region dried up. Fields bared by the plow fell victim to high winds as topsoil blew away and families took to the roads in search of a better life.

Many of these families migrated to California, a state that has suffered its own recurrent droughts. Much of California is in a state of permanent aridity – the modern-day development of the Los Angeles Basin, southern San Joaquin Valley and Imperial and Coachella valleys belies the natural desert conditions of these regions.

Even in the wetter portions of the state, dry periods are a regular part of the climate pattern. Often, years of drought are followed or preceded by years of extreme rainfall. In the Central Valley, the severe drought of 1863-1864 followed one of the wettest years which produced the Great Calamity floods of 1862. Few years fit the "normal" (average) water year.

Unlike a sudden onslaught of flood-waters, subtlety marks a drought crisis. A typical drought lasts for several years and its beginning may only be recognized in hindsight upon later analysis of precipitation patterns. One year of below-normal rainfall in itself does not necessarily equal drought. Other factors that come into play include surface runoff, soil moisture, ground-water levels and reservoir storage.

For decades, 1928-1934 stood as the state's drought of record, serving as the model to test water supply reliability for most modern northern California water projects, including the ambitious Central Valley Project.

Two great droughts in the latter part of the twentieth century shook the foundation of those plans, nature's stern reminder of the state's natural aridity. The 1976-1977 drought, although short, caused severe rationing throughout the state. It prompted some changes in common water use practices and instituted some conservation planning, but reality did not set in until a decade later when a six-year drought began in 1987.

By the time that drought ended in 1992, a host of conservation practices had become ingrained in everyday life through landscape watering restrictions, low-flow showerheads and tiered water pricing. Like many other droughts, however, it was followed by years of flooding, which served to erase from the public consciousness the overriding urge to conserve. ●

"As I have read home letters telling of 'very dry spell' …
I often compare that with this climate. There was a heavy rain here
last April, but none since. I have seen no heavy rain since early
last January. … Think and try to conceive — if possible, how dry it
must be — everything, except trees, parched and sere, watercourses
but dry beds of sand, roads six to eight inches deep of the finest dust,
soil everywhere cracked to the depth of two to six feet … indeed,
the whole surface fissured with cracks one to three inches wide."

– William H. Brewer –
Up and Down California in 1860-1864

Historic Droughts

Droughts are regional in nature. The drought of 1863-1864 had a greater impact on southern California than the rest of the state. In contrast, the drought of 1928-1934 struck the Sacramento River Basin the hardest.

During the first three water years, 1928-29, 1929-30 and 1930-31, runoff on the American River ranged from 26 to 64 percent of average. There was a temporary respite in 1932, 101 percent of average, but dry years returned in 1933, 49 percent of average runoff, and 1934, 44 percent of average.

California wasn't alone. *Business Week* reported on May 6, 1931, that "the far West faces a water shortage in 1931 that may rival the driest years in its history." The article said the April 1 snow survey revealed that the water content of the Merced River Basin snowfall was only 30 percent of normal and only 45 percent of normal throughout the rest of the Sierra Nevada. Two-thirds of the Bay-Delta's channels grew salty and the water

was unfit for irrigation. By late in the summer and fall of 1931, the Sacramento River barely kept pace with evaporation and no fresh water reached San Francisco Bay.

That drought remains the driest on record for the Sacramento River Basin. Even when compared to a 440-year period of droughts based on the study of tree growth rings, the 1928-1934 drought is the worst of the reconstructed record, which dates back to 1560.

In contrast, the more recent six-year drought that began in 1987 topped the 1928-1934 drought as the driest on record in the San Joaquin River Basin.

Other severe drought periods in California include:

1862-1864 The drought hastened the end of the ranchos; from San Joaquin County to San Diego County, thousands of cattle died, virtually ending the state's great cattle industry.

1869-1870 – Crops failed to sprout and the grass dried up in San Diego; cattle were sent to the mountains.

1877 – Severe drought in San Diego.

1895-1905 – Los Angeles' water use was greater than the local supply in the Los Angeles River and drought conditions forced rationing of water for park ponds and sprinkler systems. The drought helped build support for the Los Angeles Aqueduct.

1898-1899 – By 1899, the reservoir behind Sweetwater Dam in San Diego County was empty. This drought period extended into 1904 in some southern California locations.

1923-1924 – In all but the interior desert regions, precipitation was only 40 to 50 percent of normal and runoff in the San Joaquin and Tulare basins fell to 25 percent of normal. In southern California, the drought sparked the move to develop the Colorado River.

1943-1951 – The drought period varied: it lasted between three and four years in the central and northern Sierra Nevada and six to eight years in rest of the state. It was most severe in the central and southern coastal areas where runoff deficiencies exceeded the 1928-1934 drought. In some places, the 1951 water year ranks as the driest on record. At its maximum extent, the drought was statewide from 1947-49.

1959-1962 – Another statewide drought. It was especially dry in southern California where 1961 ranks among the driest on record.

1976-1977 – A short, but severe drought. 1977 remains the driest year on record in California.

1987-1992 – Statewide drought that fostered long-term water savings as water suppliers encouraged consumers to install low-flush toilets and low-flow shower-heads. •

1976-77 Drought

t was the driest year on record. California's runoff in 1977 was only 15 million acre-feet. Compare that to the all-time record high of 135 million acre-feet in 1983. At its lowest point, water storage in the 4.5 million acre-feet capacity Shasta Reservoir (above) stood at just 563,300 acre-feet.

On the heels of a dry 1976, the fourth driest year on record at that time, the 1977 weather conditions brought about severe water shortages throughout the state. Flows dried up in stretches of numerous waterways including the Eel, Cosumnes and Truckee rivers. Groundwater use rose drastically, resulting in severe overdraft conditions. Residents in many communities faced mandatory water rationing while everyone was urged to "Save Water" through an extensive education campaign led by the Department of Water Resources. •

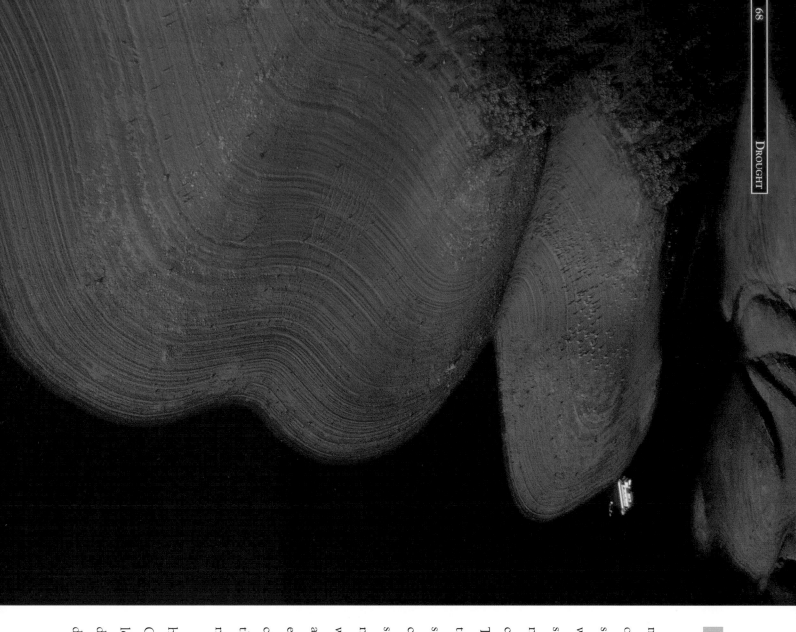

The Disconnect Between Drought and Water Use

Societal and technological advancements have allowed for uses of water once unimaginable in this semiarid state: lush landscaping, backyard pools, whirlpool bathtubs. In some ways, the successful alteration of California's natural waterscape almost belies conservation.

Turning the tap comes with such ease it has created a strange disconnect between water supply and water use. Residents are able to enjoy all the benefits of a naturally wet climate without sacrificing the weather that allows California to enjoy nicknames like the Golden State.

During times of drought, reality hits home — at least temporarily. Californians have proven they can use less water, as statistics show. Yet even during droughts, there remains the dichotomy of developing lakeside subdivisions around artificial lakes in a state whose normal climate offers little to no rainfall some six months of the year.

Not everyone reduces water use. For every resident who goes so far as to collect water after the children's baths, there is the neighbor who continues to hose off the driveway even as local officials request — or

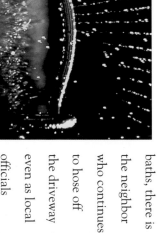

require — the use of brooms.

During the 1987-1992 drought, mandatory rationing was imposed in Santa Barbara. The city banned outright unneccesary water use: no washing cars, hosing down driveways or watering lawns. Yet the illusion of what was "natural" remained as some determined homeowners turned to a biodegradable dye to paint their lawns green. •

During the height of the 1987-1992 drought, landscaper Les Robin, above, used a biodegradable dye known as Oasis Green to paint this parched lawn in Santa Barbara.

Recreation on the state's rivers also is affected during drought.

Even winter recreation can be affected by dry conditions. During the 1976-1977 drought, many ski resorts in the Sierra Nevada were forced to reduce and even cancel their annual ski seasons as they waited in vain for snow to fall. •

With reduced instream flows, many white-water rapids disappear. Fishermen find it harder to reach the river's edge or launch a boat and as waterways become smaller and slower, water temperatures rise and fish populations suffer.

No Place to Play

Beached boats, dried-up marinas and "no swimming" signs are all striking symbols of drought. Even where resourceful resort operators install floating boat docks or extend existing boat ramps, the longer walk from parking lot to beach can cause attendance to fall. With a reduction in recreation comes a corresponding decline in income. During the short but severe 1976-1977 drought, many resort operators were forced to close their businesses and file for bankruptcy.

This family saved every drop possible during the 1987-1992 drought, using the water collected indoors on outdoor plants.

Water Cops to Rationing

Water cops are a common sight during droughts as they patrol city streets searching for gutter flooders.

Inside their homes, residents are urged to take shorter showers, wash only full loads of clothes and turn off the tap while brushing their teeth.

During droughts, most people respond to the dramatic photos of exposed shorelines and record-low reservoirs and do conserve water. State water officials observed after the 1976-1977 drought that "it is clear that Californians can carry on nearly all domestic activities, with little more

than a minor crimp in lifestyles, with a rather substantial reduction in water consumption. Few people really suffered from water shortages; they changed habits to waste less."

Whether under mandatory or voluntary conservation rules, water providers called on their customers to reduce their water use anywhere between 25 and 50 percent during the driest years of the 1987-1992 drought. The state's residents responded and even after the drought's end, water use remained lower in many metropolitan areas. •

Flush with the Urge to Conserve

During the 1987-1992 drought, water utilities offered customers rebates of several hundred dollars to replace older toilets with new ultra-low-flush models.

It was an update of the 1976-1977 drought, in which residents were encouraged to adopt the scientific theory of displacement – placing a bag or bottle filled with

water inside the toilet tank, reducing the amount of water used per flush.

By 1996, about 5 percent of California's estimated 20 million residential toilets had been replaced with ultra-low-flush varieties. These use only 1.6 gallons per flush compared to low-flush toilets (3.5 gallons per flush) and conventional toilets (6 gallons per flush). •

Drought and Nature

Cracked earth. Dying trees. Forest fires. Dry streambeds. These are all images of the natural landscape during drought. Such periodic dry spells are part of the natural cycle, and plants and animals have evolved over thousands of years to withstand such conditions. The affects of drought, nonetheless, can be devastating on the environment.

Millions of trees in the state's forests died during the 1987–1992 drought, not only from the dry conditions but from a combination of insects and disease. Weakened trees are more susceptible to the bark beetle, estimated to have killed half the trees in the Lake Tahoe area alone. Fire danger increased with each dry year and thousands of

acres were destroyed by flames. Fish were left high and dry as water levels declined and some streams literally dried up. Low water levels sparked algae blooms in some smaller lakes and reservoirs, depleting the oxygen and causing massive fish kills.

Warm water temperatures in streams and rivers killed salmon eggs. Even at fish hatcheries, water reached lethal temperatures. When coastal streams such as the Carmel River ran dry, steelhead trout could no longer return to spawn. •

"I awoke one night and thought I heard rain.
It was the dry needles of fir trees falling on the roof.
The river like some great whale lies dying in the forest."

– Barry Holstun Lopez –
River Notes: The Dance of the Herons

Drought and Agriculture

Merriam-Webster's Dictionary defines drought as "a period of dryness especially when prolonged that causes extensive damage to crops or prevents their successful growth." Historically, drought equaled disaster for California agriculture. Farmers and ranchers were wiped out as crops withered and livestock died. By virtue of the state's extensive investment in its water infrastructure, irrigated agriculture is now somewhat protected from such hydrological droughts.

Consider the 1987-1992 drought. Although farmers in some regions did suffer severe water shortages, others were able to drill wells and tap groundwater aquifers. And even as some farmers were forced to fallow their fields, others chose not to plant crops, instead selling their water to the state, which, in turn, furnished cities and wetlands with much needed supplies.

Overall, irrigated agriculture was able to withstand the drought as

farmers shifted crops, used groundwater, installed conservation systems and transferred water from farm to farm, irrigation district to irrigation district. Indeed, statewide, irrigated acreage had declined only 4 percent by the fifth year of drought.

Farmers were able to weather the state's most recent hydrological drought. What concerns agricultural representatives today, however, is the potential of what they call a "regulatory drought." From their perspective, a future drought will

be harder for agriculture to absorb because new requirements for fishery and water quality flows established post-1992 have resulted in cutbacks in years of so-called "normal" precipitation. •

"It hurts to cast into the outer darkness, to deprive of water certain tracts of land even if they have been only partially improved. To every acre that is fenced, cleared, plowed, leveled and ditched adheres the memory of backbreaking human toil; in every post hole lies buried a little of the settler's strength; every furlong of dry and crumbling ditch contains a dead hope and the ghost of strangled ambition stares out of the empty window frames of his deserted house."

– Walter V. Woehlke –

"After the Great Drouth," Sunset magazine, December 1924

The Great Projects

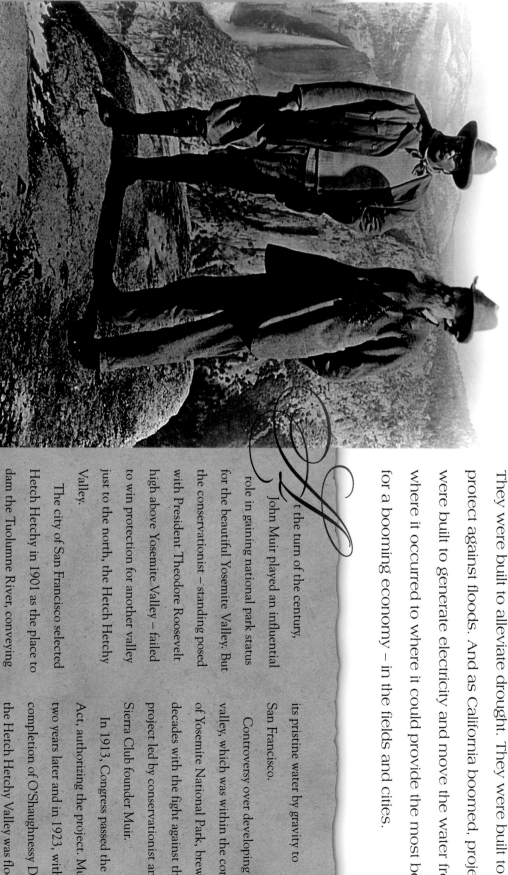

They were built to alleviate drought. They were built to protect against floods. And as California boomed, projects were built to generate electricity and move the water from where it occurred to where it could provide the most benefit for a booming economy – in the fields and cities.

At the turn of the century, John Muir played an influential role in gaining national park status for the beautiful Yosemite Valley. But the conservationist – standing posed with President Theodore Roosevelt high above Yosemite Valley – failed to win protection for another valley just to the north, the Hetch Hetchy Valley.

The city of San Francisco selected Hetch Hetchy in 1901 as the place to dam the Tuolumne River, conveying its pristine water by gravity to San Francisco.

Controversy over developing the valley, which was within the confines of Yosemite National Park, brewed for decades with the fight against the project led by conservationist and Sierra Club founder Muir.

In 1913, Congress passed the Raker Act, authorizing the project. Muir died two years later and in 1923, with completion of O'Shaughnessy Dam, the Hetch Hetchy Valley was flooded. •

"Dam Hetch Hetchy!
As well dam for water tanks
the people's cathedrals and
churches, for no holier temple
has ever been consecrated
by the heart of man."

– John Muir –
John of the Mountains

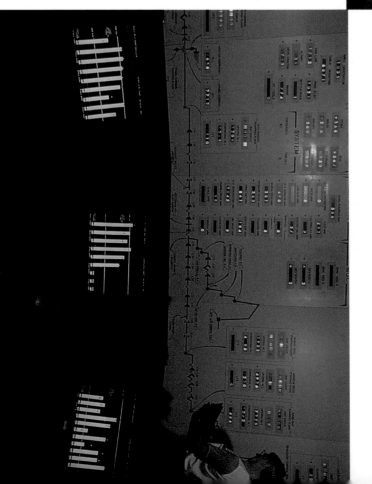

"Except for an occasional and modest reform of water institutions, most Californians felt no compelling reason to change the system. Rather, the opposite became true as they took great pride — as measured in newspaper pronouncements, popular magazines, chamber of commerce brochures and (especially) ballot-box returns — in the state's hydraulic, industrial, and agricultural achievements."

— *Norris Hundley* —
The Great Thirst

Moving the Snow

Pipes large enough for a man to stand upright. Armies of men drilling dams. The State Water Project. These projects were not without controversy, but society shared a determination that neither drought nor flood would keep them from making California great.

Californians harnessed the rivers' power to generate electricity and industry. They tapped distant waterways to bring wealth to sunny valleys. They built upstream dams and downstream levees to control flood flows. •

Aqueduct. Shasta, Friant and Folsom

through mountains and digging across deserts. Canals as wide as roads. Pumps as large as a semi truck's trailer. With a network of plumbing crisscrossing the state, Californians set about to move the snow from its natural locations to where it would best serve their needs.

The Los Angeles Aqueduct. Hetch Hetchy. The East Bay's Mokelumne River Project. The Colorado River

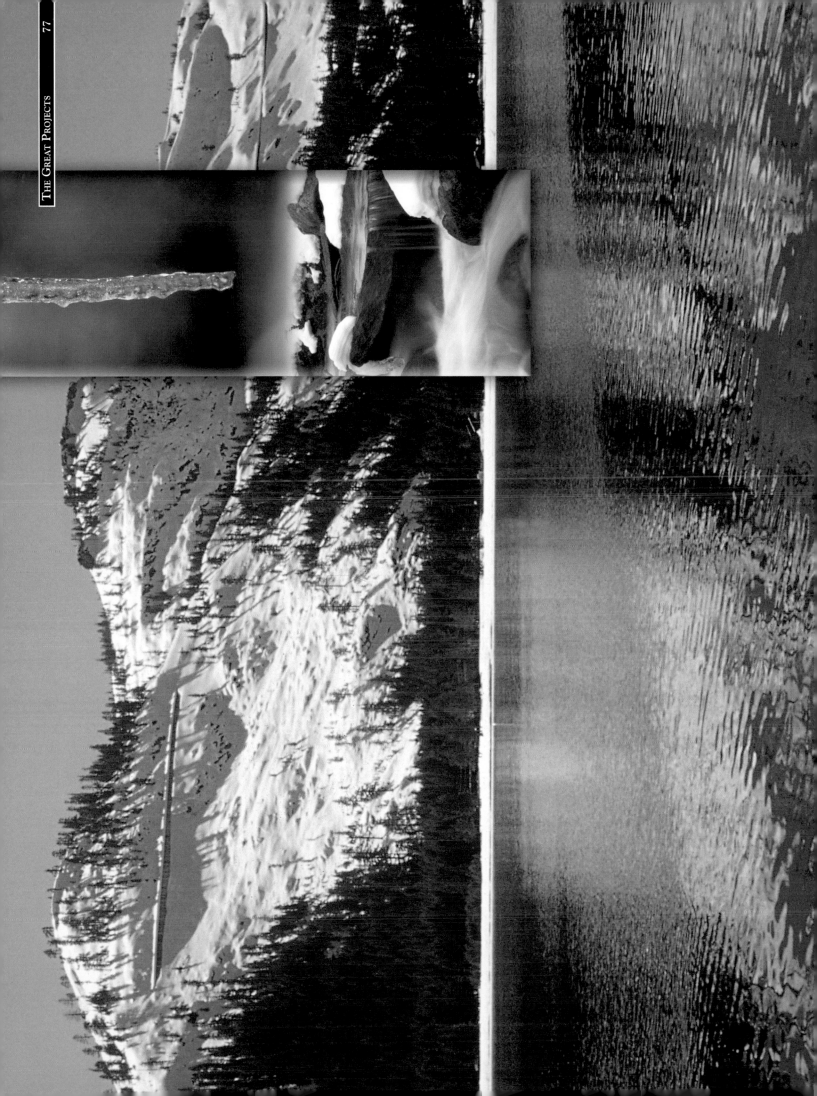

The Central Valley Project

The fear of floods and a dread of drought prompted plans to build a project to capture "surplus" water from the northern end of the Great Central Valley and move it to the drier southern end of the valley.

First envisioned in the 1870s, the Central Valley Project (CVP) did not take physical form until the 1930s and 1940s. One of the most ambitious water engineering projects constructed under the 1902 federal Reclamation Act, the CVP grew over fifty years to become one of the largest water storage and transport systems in the world.

Early components of the CVP provided construction jobs during the Great Depression, and the waters from this system helped establish the Central Valley as the richest agricultural region in the world.

"Every drop of water that runs to the sea without rendering a commercial return is a public waste."

— Herbert Hoover —

Today, in addition to supplying vitally needed irrigation water to the Sacramento and San Joaquin valleys, the CVP also provides water to cities and industries in Sacramento and the east and south Bay Areas and to fish hatcheries and wildlife refuges throughout the Central Valley.

In the twenty-first century, the seemingly straightforward goals of harnessing water for economic prosperity and protecting development from floods have given way to a complex, multiuse/multi-interest world in which man's progress must somehow be balanced against the needs of nature. When individual's futures and town economies rely on historical water rights, however, it is difficult to incorporate a new approach – no matter how widely supported or legally mandated. •

Shasta dam site, prior to construction.

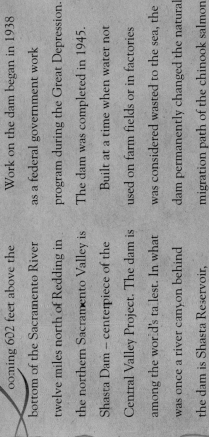

Work on the dam began in 1938 as a federal government work program during the Great Depression. The dam was completed in 1945.

Built at a time when water not used on farm fields or in factories was considered wasted to the sea, the dam permanently changed the natural migration path of the chinook salmon. The Sacramento River is home to four runs of chinook salmon, which were cut off from their native spawning grounds above the dam. In 1989 the winter-run chinook was declared an endangered species, requiring a change in traditional dam operations to protect and restore the fish. •

ooming 602 feet above the bottom of the Sacramento River twelve miles north of Redding in the northern Sacramento Valley is Shasta Dam – centerpiece of the Central Valley Project. The dam is among the world's tallest. In what was once a river canyon behind the dam is Shasta Reservoir, capable of storing up to 4.5 million acre-feet of water.

The dam captures the flow of the Pit, McCloud and upper Sacramento rivers, protecting the valley from ferocious flooding and storing water for use in more arid regions and seasons.

"We had the river licked.
Pinned down, shoulders right on the mat.
Hell, that's what we came up here for."

– Francis Trenholm Crowe –
Supervisor of Shasta Dam construction

An excerpt from Joan Didion's *The White Album*

Since the afternoon in 1967 when I first saw Hoover Dam, its image has never been entirely absent from my inner eye. I will be talking to someone in Los Angeles, say, or New York, and suddenly the dam will materialize, its pristine concave face gleaming white against the harsh rusts and taupes and mauves of that rock canyon hundreds or thousands of miles from where I am. . . . Frequently I wonder what is happening at the dam this instant, at this precise intersection of time and space, how much water is being released to fill downstream orders and what lights are flashing and which generators are in full use and which just spinning free. . . .

Of course the dam derives some of its emotional effect from precisely that aspect, that sense of being a monument to a faith since misplaced. "They died to make the desert bloom," reads a plaque dedicated to the 96 men who died building this first of the great high dams, and in context the worn phrase touches, suggests all of that trust in harnessing resources, in the meliorative power of the dynamo, so central to the early Thirties. . . .

But history does not explain it all, does not entirely suggest what makes that dam so affecting. Nor, even, does energy, the massive involvement with power and pressure and the transparent sexual overtones to that involvement. Once when I revisited the dam I walked through it with a man from the Bureau of Reclamation. . . . Once in a while he would explain something, usually in that recondite language having to do with "peaking

power," with "outages" and "dewatering," but on the whole we spent the afternoon in a world so alien, so complete and so beautiful unto itself that it was scarcely necessary to speak at all. We saw almost no one. Cranes moved above us as if under their own volition. Generators roared. Transformers hummed. The gratings on which we stood vibrated. We watched a hundred-ton steel shaft plunging down to that place where the water was. And finally we got down to that place where the water was, where the water sucked out of Lake Mead roared through thirty-foot penstocks and then into thirteen-foot penstocks and finally into the turbines themselves. "Touch it," the Reclamation said, and I did, and for a long time I just stood there with my hands on the turbine. It was a

peculiar moment, but so explicit as to suggest nothing beyond itself. . . .

I walked across the marble star map that traces a sidereal revolution of the equinox and fixes forever, the Reclamation man had told me, for all time and for all people who can read the stars, the date the dam was dedicated. The star map was, he had said, for when we were all gone and the dam was left. I had not thought much of it when he said it, but I thought of it then, with the wind whining and the sun dropping behind a mesa with the finality of a sunset in space. Of course that was the image I had seen always, seen it without quite realizing what I saw, a dynamo finally free of man, splendid at last in its absolute isolation, transmitting power and releasing water to a world where no one is.

ohn Wesley Powell was the first man to successfully navigate the Colorado River. A one-armed Civil War major, Powell led three boats through the Grand Canyon in 1869, proving that the trip could be made.

It was a turning point in the Colorado River Basin, one of the last areas in the United States to be explored by Anglo-Americans. As late as the 1850s the basin appeared on U.S. maps as a 500- by 200-mile region marked "unexplored."

After surveying the Southwestern United States from 1867 to 1872, Powell concluded that Western settlement would be successful only if land and water rights were inseparable and the existing land and water monopolies were broken up. Water rights must be used or forfeited, he maintained, and the federal government must step in to control the rivers, irrigate the lands and equitably distribute the West's most precious resource, water.

These views, contained in his 1878 book, The Report on the Arid Region of the United States, challenged the predominant Western land-use policies of the 1800s. In the book, Powell made three recommendations: sell no more farmland that doesn't have access to water; set property boundaries to encompass natural watersheds to avoid competition for streams; and do not rely on private water companies to develop water projects.

Although Powell's ideas were generally rejected at the time, the essence of his recommendations was later incorporated into the federal reclamation program established by Congress in 1902. •

Grand Plans and Great Vision

The turbulent Colorado River shaped the natural Southwest, cutting through the sandstone of the plateau to form a rust colored landscape of barren vistas and deep canyons. By harnessing the river's valuable resource – water – man reshaped the arid Southwest.

Although the Anasazi Indians developed a Colorado River fed distribution system in 600 A.D. in New Mexico and early farmers tapped the river in the 1800s in Utah and California, it was Boulder Dam – later renamed Hoover Dam – that initiated the river's modern-day development.

The Colorado River Aqueduct

"Don't miss seeing the building of Boulder Dam.

It's the biggest thing that's ever been done with water since

Noah made the flood look foolish. You know how big the

Grand Canyon is. Well, they just stop up one end of it, and

make the water come out through a drinking fountain.

They have only been bothered with two things: one is silt and the

other is Senatorial investigations. They both clog everything up."

— Will Rogers –
"The People's Jester"

Legislation to build such a project was first introduced in 1922 by two California congressmen. The proposal came during an era of grand plans and great vision, when water flowing down a stream was considered wasted unless it was used by man, when the goal was to use the technology of the day to improve the human condition, when a developing nation still was eager to settle the West.

Ten years later, the dam began to take shape after years of conflict among the seven states that share the Colorado River had been settled through negotiations led by Herbert Hoover, then serving as secretary of commerce.

By the time an army of men began pouring concrete in Black Canyon to form the dam's curved, concave face that rises 726 feet from the canyon floor, the Great Depression era of federal public works programs had begun. Renowned as one of the great engineering feats of the century, Hoover Dam was completed in 1935.

Today, Hoover Dam is one of forty-nine major water facilities in the seven-state Colorado River Basin that give the river the title of the "most appropriated" stream in the world. Its 28.5 million acre-foot reservoir, Lake Mead, is the largest man-made reservoir in the United States. •

The first surveys to determine the best route by which Los Angeles could reach across the desert and tap the Colorado River were conducted in 1923 by none other than William Mulholland. It had been a decade since completion of his famous Los Angeles Aqueduct and the city's population had reached 1 million when drought struck. Reliable water to ensure continued prosperity was the goal and the Colorado the chosen source.

More than a hundred routes were surveyed, mostly by foot, with some use of horses, mule pack trains, boats and, later, Model-Ts and Model-As. According to early surveyors, the river "was the color of hot chocolate" with the same consistency. During the earliest surveys Boulder (Hoover) Dam had not yet been authorized, so devising a way to remove silt from the water before it went into the aqueduct was a major concern.

Other south coast cities joined with Los Angeles and in 1928 formed the Metropolitan Water District of Southern California (MWD).

The route was selected in 1931 and seven months later, at the height of the Great Depression, voters in southern California's coastal counties went to the polls and approved a $220 million bond measure to build the Colorado River Aqueduct. Construction began in 1933, two years after work started on Hoover Dam. The 242-mile aqueduct was completed in 1939 and MWD's distribution system was completed in June 1941.

At first, there was little demand for Colorado River Aqueduct water, but Pearl Harbor and America's entry into World War II changed that as southern California became the headquarters for the nation's defense industry. •

"We have tried to make the arid West into
what it was never meant to be and cannot remain,
the Garden of the World and the home of multiple millions. …
The West — the habitable parts of it — is a splendid habitat
for a limited population living within the country's
rules of sparseness and mobility."

— Wallace Stegner —
The American West as Living Space

The California Aqueduct.

The Los Angeles Aqueduct is one component of the man-made waterway systems that crisscross the length of modern-day California. The system relies on gravity to convey water 233 miles from the eastern Sierra Nevada to the city of Los Angeles. Built in the early twentieth century, the aqueduct's construction was a mammoth undertaking. Even before a work force of 5,000 began construction of the aqueduct in 1907, 500 miles of paved roads and rails, 240 miles of telephone wire and more than 2,300 buildings had to be built along the aqueduct's route. For electricity, two hydropower plants were constructed to power the dredges, excavators and drills, and to light the camps and tunnels.

At the time of its design, the aqueduct could deliver four times more water than Los Angeles needed and allowed the city to become the urban center of 3.7 million people it is today.

On November 5, 1913, 43,000 spectators gathered in the San Fernando Valley to watch the first stream of pure Sierra Nevada water arrive. "There it is – take it," William Mullholland shouted to the crowd when the water appeared.

In 1940, the aqueduct was extended 105 miles north to tap the waters of the Mono Basin. A second aqueduct to carry more water was completed in 1970. •

Building the Los Angeles Aqueduct.

Artificial Rivers

Stretching 444 miles from north to south, the California Aqueduct (left) delivers 3 million acre-feet of water a year to farmers in the San Joaquin Valley, and cities in the south Bay area, coastal California and southern California.

The aqueduct is a vital component of the State Water Project (SWP), which begins on the Feather River in the northern Sacramento Valley. Traveling through the San Joaquin Valley, the aqueduct is readily visible from another key north-south link – Interstate 5. The aqueduct and the pump station at Edmonston Pumping Plant that lifts the water over the Tehachapi Mountains are engineering marvels that attract visitors from all over the world.

Given the historic animosity between northern and southern Californians over water, the aqueduct is a physical reminder that the two different regions are one state. Championed by Gov. Edmund G. "Pat" Brown, construction of the SWP was funded by a statewide bond measure passed in 1960 – the largest water bond in state history. At one point in his life, the late Gov. Brown is reported to have explained his reason for supporting the measure by noting facetiously that he "didn't want all the people in the southland to move north." •

arge or small, waterworks projects were no cause for celebration in semiarid California.

Consider Oroville Dam. At the June 1, 1957, dedication ceremony, a special train transported visitors from as far away as southern California to the dam site north of Sacramento, where Gov. Goodwin Knight espoused the benefits of a project linking northern and southern California. "In spite of its proportions which stagger the imagination, the project is merely the beginning of greater things to come," Knight said. The train then carried celebrants back to Sacramento for a parade, barbecue and entertainment. This event marked only the relocation of the railroad tracks at the site of the dam; it would be another five years before work would actually began on the dam.

Smaller projects were no less celebrated. At the April 19, 1888, dedication of the new Sweetwater Dam in National City, two pipes with 1-inch nozzles sprayed water 70 feet into the air. More than 3,000 people – including men and women dressed in their finest – turned out to celebrate the completion of the 90-foot stone and concrete dam. Designed to deliver water to nearby suburbs, the dam was heralded as the region's gateway to prosperity. "While five years ago the town was dead, it is now ringing with the hammer," Judge Puterbaugh said at the event. "Water is king." ●

Folsom Dam dedication in 1956.

"William Mullholland had been

taken to the top of the mountain;

and like Moses he had been granted a vision

of his people's deliverance. His miracle was

not to part the sea, but to part the sands;

not to keep the waters back,

but to bring them forth and create

rivers in the desert."

– Margaret Leslie Davis –
Rivers in the Desert

or William Mullholland, if the Los Angeles Aqueduct ceremony was the high point, St. Francis Dam was the low point. Construction of St. Francis Dam in the San Francisquito Canyon some 20 miles north of the San Fernando Valley (near today's city of Santa Clarita) began in 1926. The great arch-shaped structure was completed in 1928, forming a 34,000 acre-feet reservoir that served as emergency storage and recaptured aqueduct water released by the valley's power plants.

But before the end of 1927, the dam's abutments had soaked enough to swell, and by January 1928 two cracks appeared on the face of the dam. As water leaked through these seams, an inspection on March 12, 1928, led Mullholland to determine that the structure appeared safe. Later that night, shortly before midnight, the dam collapsed, sending a wall of water down San Francisquito Canyon, ripping out trees and destroying a settlement for dam employees.

The flow then turned into the Santa Clara River and raced along at 18 miles an hour toward the Pacific Ocean. In its wake it carried uprooted trees, houses and other debris. A 60-foot-tall wave washed through Castaic Junction, where only a handful of people survived.

The flood roared through the towns of Piru and Fillmore, reaching Santa Paula at 3:30 in the morning, four and a half hours after the dam first broke. Still, the wall of water was 25 feet tall.

Altogether, some 450 people were killed, many having had no warning and little chance to escape. Five separate investigations reached the same conclusion: while the concrete structure itself was sound, the hillside had collapsed. Mullholland, crushed by the tragedy, accepted full responsibility and soon retired as Los Angeles' chief water engineer. The collapse led to creation of a state dam safety program. •

Distributing the Resource – California United

To cross the Tehachapi Mountains into southern California, water is elevated nearly 2,000 feet – more water pumped higher than anywhere else in the world – at A.D. Edmonston Pumping Plant.

It is perhaps here that the saying "water flows downhill, except toward money," is most apropos. For on the other side of the mountains lies the Los Angeles Basin, home to 9.5 million people.

Part of the State Water Project, construction of the plant began in 1965. On October 7, 1971, Gov. Ronald Reagan dedicated the newly completed pumping plant. Several thousand people turned out for the dedication. "For the first time in history, California will be united – north and south – with a water transportation system that truly distributes one of the state's most important resources," Reagan said.

He gave the command for the pumps to start and thirty minutes later

officials reported seeing the water come out the other side. The completion of Edmonston Pumping Plant was heralded in a *San Diego Union* editorial that noted: "The first flow of Feather River water into southern California today represents a triumph over an environmental problem – the arid climate that threatened to limit the future of a vast area of our state.

"Surely the same resources of technology and engineering that went into the triumph can meet the challenge of harnessing water resources for the public benefit with a minimum of disturbance to the natural environment."

The plant is named for Arthur D. Edmonston, who served as state engineer and chief of the division of water resources from 1950-1955. He also directed the early planning of the Central Valley Project, State Water Project and State Water Plan. ●

n August 18, 1962, President John F. Kennedy and California Gov. Edmund G. "Pat" Brown pushed the plungers, detonating the dynamite and breaking ground for the new federal-state off-stream San Luis Reservoir located in the barren hills 70 miles south of the Delta. Thousands of spectators were on hand at the ceremony.

Kennedy praised Brown for his success in winning approval of the State Water Project.

"For many years some believed that the water problems of this state were too controversial and too complicated to solve," Kennedy said. "They believed that there was no escaping the effects of drought and flood, the shortages which crippled industrial development and the staggering waste of soil and water which future Californians would require. Many state administrations in California wrestled with this problem and some gains were made, but I believe that all Californians will long remember the water leadership of Gov. Pat Brown." ●

California's Transformed Waterscape

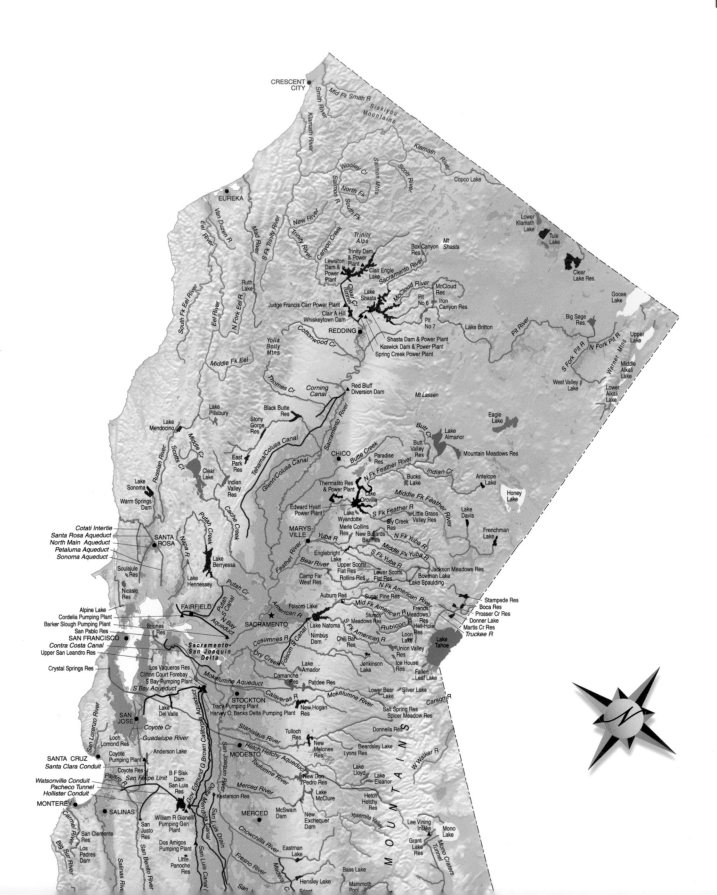

CRESCENT CITY

Mid Fk Smith R

Smith River

Siskiyou Mountains

Klamath River

Wooley Cr

Salmon Mtns

Scott River

Klamath River

Copco Lake

Lower Klamath Lake

Tule Lake

EUREKA

North Fk

Salmon R

South Fk

Mt Shasta

Clear Lake Res.

New River

Van Duzen R

Mad River

S Fk Trinity River

Trinity River

Canyon Creek

Trinity Alps

Box Canyon Res

Eel River

Trinity Dam & Power Plant

Lewiston Dam & Power Plant

Clair Engle Lake

Sacramento River

McCloud River

McCloud Res

Goose Lake

Ruth Lake

N Fork Eel R

Clear Cr Tunnel

Lake Shasta

Pit No 6

Iron Canyon Res

Big Sage Res.

South Fk Eel River

Judge Francis Carr Power Plant

Clair A Hill Whiskeytown Dam

Pit No 7

Lake Britton

Pit River

S Fork Pit R

N Fork Pit R

Upper Lake

Middle Fk Eel

Yolla Bolly Mtns

Cottonwood Cr

REDDING

Shasta Dam & Power Plant

Keswick Dam & Power Plant

Spring Creek Power Plant

Warner Mtns

Middle Alkali Lake

West Valley Lake

Lower Alkali Lake

Thomes Cr

Corning Canal

Red Bluff Diversion Dam

Mt Lassen

Lake Pillsbury

Black Butte Res

Eagle Lake

Lake Mendocino

Stony Gorge Res

Sacramento River

Butt Cr

Lake Almanor

Mountain Meadows Res

Middle Fk Eel

Tehama/Colusa Canal

East Park Res

CHICO

Butte Creek

Butt Valley Res

Scotts Cr

Clear Lake

Paradise Res

N Fk Feather River

Indian Cr

Antelope Lake

Lake Sonoma

Russian River

Middle Cr

Indian Valley Res

Glenn/Colusa Canal

Bucks Lake

Honey Lake

Warm Springs Dam

Cache Creek

Thermalito Res & Power Plant

Lake Oroville

Middle Fk Feather River

Lake Davis

Cotati Intertie

Santa Rosa Aqueduct

North Main Aqueduct

Petaluma Aqueduct

Sonoma Aqueduct

SANTA ROSA

Putah Creek

Napa R

Edward Hyatt Power Plant

Lake Wyandotte

S Fk Feather R

Little Grass Valley Res

Frenchman Lake

MARYS-VILLE

Merle Collins

Sly Creek Res

N Fk Yuba R

Soulajule Res

Lake Berryessa

Yuba R

New Bullards Bar Res

Middle Fk Yuba R

Nicasio Res

Lake Hennessey

Feather River

Englebright Lake

Bear River

S Fk Yuba R

Upper Scotts Flat Res

Jackson Meadows Res

Alpine Lake

Putah Cr

Camp Far West Res

Lower Scotts Flat Res

Bowman Lake

Lake Spaulding

Cordelia Pumping Plant

FAIRFIELD

Putah Canal

Rollins Res

Barker Slough Pumping Plant

N Bay Aqueduct

Auburn Res

Sugar Pine Res

N Fk American River

Stampede Res

San Pablo Res

Briones Res

Folsom Lake

Mid Fk American R

French Meadows Res

Boca Res

Prosser Cr Res

SAN FRANCISCO

American R

SACRAMENTO

Lake Natoma

S Fk American R

Rubicon R

Hell-Hole Res

Donner Lake

Martis Cr Res

Contra Costa Canal

Folsom Canal

Nimbus Dam

Chili Bar Res

Truckee R

Upper San Leandro Res

Sacramento-San Joaquin Delta

Cosumnes R

Loon Lake

Lake Tahoe

Crystal Springs Res

Dry Creek

Lake Amador

Jenkinson Lake

Union Valley Res

Ice House Res

Los Vaqueros Res

Clifton Court Forebay

S Bay Pumping Plant

S Bay Aqueduct

Mokelumne Aqueduct

Camanche Res

Pardee Res

Lower Bear Lake

Silver Lake

Fallen Leaf Lake

Calaveras R

Mokelumne River

Salt Spring Res

Carson R

SAN JOSE

Lake Del Valle

STOCKTON

Tracy Pumping Plant

Harvey O. Banks Delta Pumping Plant

New Hogan Res

Spicer Meadow Res

Coyote Cr

Stanislaus River

Donnells Res

Loch Lomond Res

Guadalupe River

Tulloch Res

Beardsley Lake

W Walker R

SANTA CRUZ

Santa Clara Conduit

Coyote Pumping Plant

Anderson Lake

MODESTO

New Melones Res

Lyons Res

San Joaquin River

Lake Lloyd

Watsonville Conduit

Pacheco Tunnel

Hollister Conduit

Coyote Res

San Felipe Unit

B F Sisk Dam

San Luis Res

Hetch Hetchy Aqueduct

Toulumne River

New Don Pedro Res

Lake Eleanor

Lake Lloyd

Pajaro R

Delta Mendota Canal

Hetch Hetchy Res

MONTEREY

Merced River

Kesterson Res

San Luis Drain

McSwain Dam

MERCED

New Exchequer Dam

Yosemite Valley

Lee Vining Intake

Mono Lake

SALINAS

Carmel River

William R Gianelli Pumping Gen Plant

Chowchilla River

Grant Lake Res

Mono Craters Tunnel

San Clemente Res

San Justo Res

Dos Amigos Pumping Plant

Eastman Lake

Salinas River

San Benito River

San Luis Canal

Little Panoche Res

Fresno River

Bass Lake

Los Padres Dam

Big Sur River

San Luis Canal

Hensley Lake

Mammoth Pool

SAN JOSE

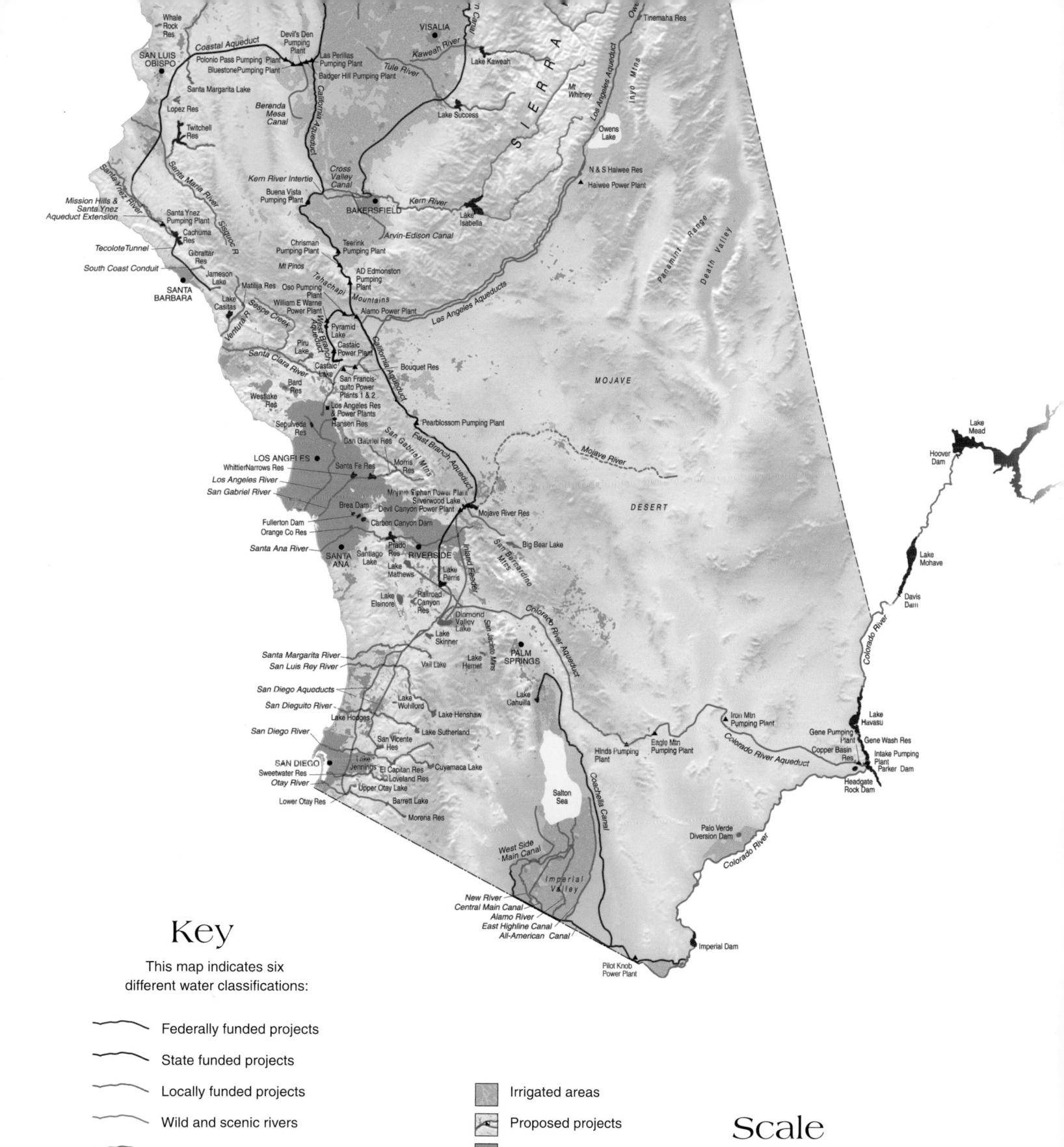

Whale Rock Res
Coastal Aqueduct
Devil's Den Pumping Plant
VISALIA
Tinemaha Res
SAN LUIS OBISPO
Polonio Pass Pumping Plant
BluestonePumping Plant
Las Perlilas Pumping Plant
Kaweah River
Lake Kaweah
Owens Lake
Tule River
Santa Margarita Lake
California Aqueduct
Badger Hill Pumping Plant
SIERRA
Mt Whitney
Los Angeles Aqueduct
Lopez Res
Beranda Mesa Canal
Lake Success
Inyo Mtns
Twitchell Res
Cross Valley Canal
N & S Haiwee Res
Santa Maria River
Santa Ynez Pumping Plant
Kern River Intertie
Buena Vista Pumping Plant
Kern River
Haiwee Power Plant
Mission Hills & Santa Ynez Aqueduct Extension
Cachuma Res
BAKERSFIELD
Lake Isabella
Panamint Range
Death Valley
Tecolote Tunnel
Gibraltar Res
Arvin-Edison Canal
Santa Ynez River
Sisquoc R
Jameson Lake
Chrisman Pumping Plant
Teerink Pumping Plant
SANTA BARBARA
Lake Casitas
Matilija Res
Mt Pinos
Oso Pumping Plant
AD Edmonston Pumping Plant
South Coast Conduit
Tehachapi Mountains
William E Warne Power Plant
Alamo Power Plant
Los Angeles Aqueducts
MOJAVE
Ventura R
Sespe Creek
Pyramid Lake
Castaic Power Plant
Piru Lake
Castaic Lake
Bouquet Res
Santa Clara River
West Branch Aqueduct
San Francis-quito Power Plants 1 & 2
Bard Res
Los Angeles Res & Power Plants
Westlake Res
Sepulveda Res
Hansen Res
East Branch Aqueduct
Pearblossom Pumping Plant
DESERT
San Gabriel Mtns
Mojave River
LOS ANGELES
WhittierNarrows Res
San Gabriel Res
Morris Res
Los Angeles River
Santa Fe Res
San Gabriel River
Mojave Siphon Power Plant
Silverwood Lake
Mojave River Res
Brea Dam
Devil Canyon Power Plant
Fullerton Dam
Carbon Canyon Dam
Orange Co Res
Big Bear Lake
Santa Ana River
Prado Res
SANTA ANA
Santiago Lake
RIVERSIDE
San Bernardino Mtns
Lake Mathews
Lake Perris
Inland Feeder
Lake Elsinore
Railroad Canyon Res
Diamond Valley Lake
San Jacinto Mtns
Colorado River Aqueduct
Lake Skinner
Santa Margarita River
Vail Lake
PALM SPRINGS
San Luis Rey River
Lake Hemet
San Diego Aqueducts
Lake Wohlford
Lake Cahuilla
San Dieguito River
Lake Henshaw
Iron Mtn Pumping Plant
Lake Hodges
Lake Sutherland
Eagle Mtn Pumping Plant
Lake Havasu
San Diego River
San Vicente Res
Hinds Pumping Plant
Colorado River Aqueduct
Gene Pumping Plant
Gene Wash Res
SAN DIEGO
Lake Jennings
El Capitan Res
Cuyamaca Lake
Copper Basin Res
Intake Pumping Plant
Sweetwater Res
Loveland Res
Parker Dam
Otay River
Upper Otay Lake
Salton Sea
Headgate Rock Dam
Lower Otay Res
Barrett Lake
Morena Res
Coachella Canal
Palo Verde Diversion Dam
Colorado River
West Side Main Canal
Imperial Valley
New River
Central Main Canal
Alamo River
East Highline Canal
All-American Canal
Imperial Dam
Pilot Knob Power Plant

Lake Mead
Hoover Dam
Lake Mohave
Davis Dam
Colorado River

Key

This map indicates six different water classifications:

Federally funded projects

State funded projects

Locally funded projects

Wild and scenic rivers

Naturally occurring lakes, rivers, etc.

Saline or alkaline lakes

Irrigated areas

Proposed projects

Urbanized areas

Pumping or power plants

Scale

0 10 20 30 40 50 60 70 80 mi.

0   20   40   60   80   100 km.

EXCESSIVE

REVERENCE?

"Some of us who live in arid parts of the world think about water with a reverence others might find excessive. The water I will draw tomorrow from my tap in Malibu is today crossing the Mojave Desert from the Colorado River, and I will draw tomorrow from my tap in Malibu is today crossing the Mojave Desert from the Colorado River, and I like to think about exactly where that water is.

The water I will drink tonight in a restaurant in Hollywood is by now well down the Los Angeles Aqueduct from the Owens River, and I also think about exactly where that water is: I particularly like to imagine it as it cascades down the 45-degree stone steps that aerate Owens water after its airless passage through the mountain pipes and siphons.

As it happens my own reverence for water has always taken the form of this constant meditation upon where the water is, of an obsessive interest not in the politics of water but in the waterworks themselves, in the movement of water through aqueducts and siphons and pumps and forebays and afterbays and weirs and drains, in plumbing on the grand scale."

— Joan Didion —
The White Album

Fertile Valleys

Salinas. Sacramento. San Joaquin. Imperial. Napa.

These are the valleys that feed the world, that produce lettuce, rice, winter vegetables and world-class grapes and wine.

California is the nation's leading farm state. Although its 8 million irrigated acres represent just 3 percent of the total farmland, more than half of the United States' fruits, nuts and vegetables are grown in California.

There are three key reasons for such productivity: rich, fertile soil deposited by the seasonal overflow of rivers; a climate with long, dry summers, in which irrigation water can be applied at just the right time, in just the right amount; and the development of irrigation systems to tap groundwater aquifers and surface streams.

Fortune seekers who failed in their quest to find wealth in the foothills discovered riches in the transformation of the thirty-first state into a prolific farmland. These pioneers looked beyond the sagebrush and dry, golden grass of summer – even the desert sands of the Imperial Valley – and envisioned green, productive fields. The only thing they needed was water.

For property adjacent to streams and rivers, water was within easy reach. For vast amounts of land, however, irrigation required ingenuity and engineering. Some farmers turned to distant surface water sources, such as the Colorado River. Others tapped groundwater aquifers beneath their feet, using the force of the wind to pump water to the surface of the fertile land.

Ultimately, what was first seen as a hindrance – little to no natural precipitation during much of the year – became a benefit: artificial irrigation allows the farmer to control how much and when a crop is watered. •

"... the grandest and most telling of California landscapes is outspread before you. At your feet lies the great Central Valley glowing golden in the sunshine, extending north and south farther than the eye can reach, one smooth, flowery, lake-like bed of fertile soil."

– John Muir –
The Mountains of California

"In the summer the entire country was a wavy wheat-field from one extreme to the other. To the wayfarer as he journeyed along its dusty roads and traveled along its well-kept farms wherein the ancient hospitality still found lodgement, the vibrating fields, animated by the gentle northern breezes, resplendent in the varied tints of the growing sun, gave a rich carpetry to Mother Earth that was charming to the eye."

– Sol P. Elias –
Stories of the Stanislaus

rrigated agriculture has a long history in California. Irrigated orchards and vineyards were common at Spanish missions.

Yet the state's first claim to farming fame was as a leading producer of dryland wheat. In the 1850s, farmers discovered California's climate was ideal for wheat production because its long, dry summers prolonged the harvest and made the wheat dry and hard. Wheat required little initial capital outlay.

Wheat could be transported long distances without fear of spoilage and it provided excellent ballast for ships sailing eastward from San Francisco.

Ultimately, the wheat yield-per-acre average in the Sacramento and San Joaquin valleys surpassed that of the Midwest. Production climbed from 17,000 bushels in 1850 to 16 million by 1873 and three years later, California was the nation's leader in wheat production. Wheat growing reached its peak in the early 1890s with production of 40 million bushels. ●

" ... Two miles from the heart of our city a man could come to the desert and feel the loneliness of a desolate area, a place lost in the earth, far from the solace of human thought. Standing at the edge of our city, a man could feel that we had made this place of streets and dwellings in the stillness and loneliness of the desert, and that we had done a brave thing. We had come to this dry area that was without history, and we had paused in it and built our houses and we were slowly creating the legend of our life. We were digging for water and we were leading streams through the dry land. We were planting and ploughing and standing in the midst of the garden we were making."

– William Saroyan –
Fresno Stories

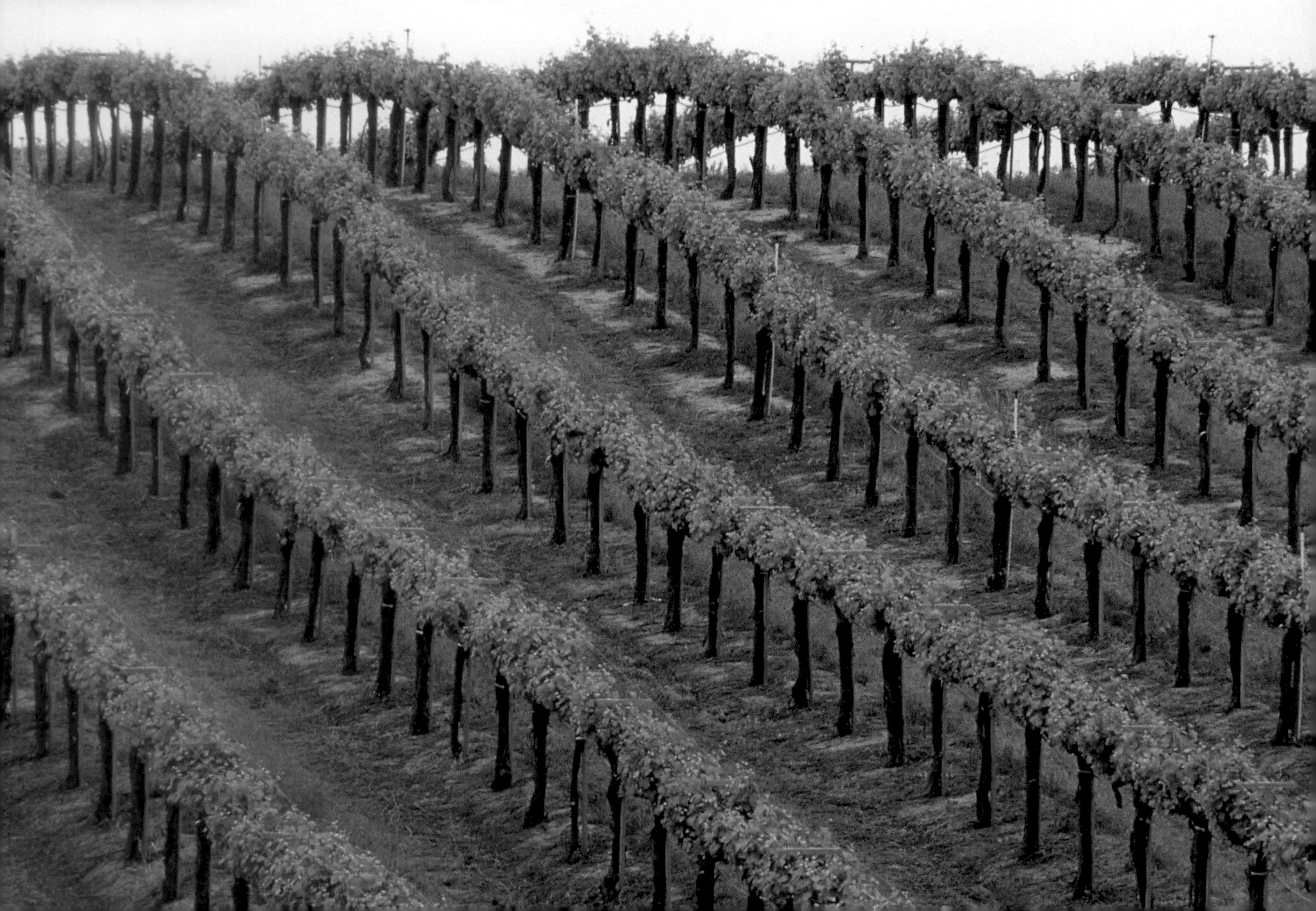

"In 1884 Riverside had the largest acreage of vines and trees of any of the colonies giving attention to orange and raisin culture south of the Sierra Madre. Yet no farther back than 1870, this valley, now so smiling and yielding such lavish returns to its cultivators, was but a silent waste, mantled in Spring-time with gay flowers and tall wild grasses. The soil is composed largely of disintegrated rock, washed from the surrounding mountains by the storms of ages, and possesses almost boundless powers of production. But these powers were dormant. Something was needed to arouse them, and that something was simply the voice of running water."

– Emma H. Adams –

To and Fro in Southern California, 1887

For the rest of the nation, California's unique climate made it a natural sensation for those living east of the Rocky Mountains. Where else could you find an orange grove nestled in a warm, fertile valley with a snow-capped mountain looming in the background?

Before the invention of the refrigerated railcar, the only markets for perishable produce were local. Experiments on such a car began in the 1860s and by 1872, East Coast states were shipping meat by rail. The idea of using the so-called reefer cars for fruit and vegetables quickly caught on and in 1887, California shipments of citrus reached 2,200 loads. By 1900, an orange grown in sunny California could be on a grocer's shelf in New York in two to three weeks. Access to these new markets encouraged California's agriculture to expand and California to become the nation's No. 1 farm state. ●

Irrigation Techniques

There are three common irrigation methods used in California, furrow or flood, sprinkler and drip. A farmer's choice of irrigation method is determined primarily by three factors: the type of soil to be irrigated, the topo-

graphy of the land, and the crop to be grown. In level valleys, furrow or flood irrigation has been the traditional choice, although many growers are switching to drip systems to conserve water. Hilly areas require the use of sprinklers or a drip irrigation system.

In a common furrow system, narrow dirt ditches spanning the length of the field divert water from a head ditch running across the upper end of the field. A ditch at the bottom of the field typically catches whatever water runs off the land for use on another field. (This water is known as tail water.) In a furrow system, the soil itself is used to convey the water. To improve the uniform distribution of water, many farmers use a high-tech laser to regrade the land and create a slight, uniform slope down the field. This laser leveling process reduces problems with low and high spots, reducing the amount of water needed to irrigate.

Flood irrigation follows the same principle, although the fields typically have no furrows.

Sprinklers and surface drip systems can apply water more slowly and accurately, giving growers more flexibility to apply just the amount of water needed for optimum plant growth. In some cases, drip lines are buried to deliver water right to the root zones. When drip emitters are used at the base of the plant they not only save

water, but decrease weed growth, improve yields and reduce the amount of fertilizer and other chemicals needed. Because drip systems are the most expensive, they are more commonly used on higher value crops such as strawberries and grapes.

In an effort to conserve water, farmers also use sophisticated technology to determine the amount of water a plant needs and the optimal time to provide this water. •

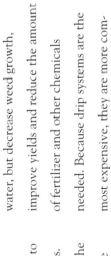

"I want to be with people who
submerge in the task,
who go into the fields to harvest and work in a row
and pass the bags along,
who stand in the line and haul in their places,
who are not parlor generals and field deserters
but move in a common rhythm
when the food must come in or
the fire be put out."

– Marge Piercy –
Circles on the Water

Golden Boy
Tomatoes

Early Girl
Tomatoes

Homegrown
&
Vine Ripened

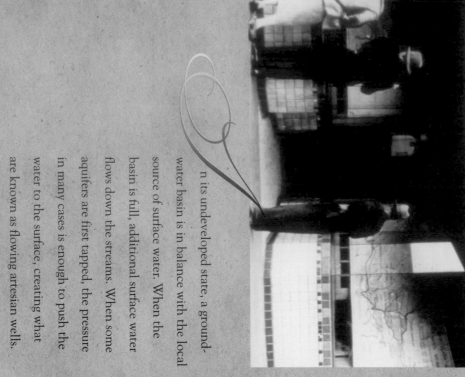

William H. Brewer described a scene in the Santa Clara Valley in 1861: "Formerly there was a lack of water here. ... The large streams that run into the valley either sink or dwindle away to mere ponds. So hundreds of artesian wells have been bored. Sometimes water is struck within a hundred feet, but many wells are three or four hundred feet deep. Many of these overflow, often with a large stream, but with the majority the water only rises near the surface without overflowing. It is then pumped up by windmills, and hundreds of these may be seen in motion every afternoon when there is a strong breeze."

By 1865, there were close to 500 wells in the valley as settlers switched from dryland farming to irrigated agriculture. As agriculture expanded, so did the number of wells and the amount of water pumped from below ground. Wells were drilled deeper and deeper as the water table dropped. The heavy groundwater pumping and several periods of years of below-

n its undeveloped state, a groundwater basin is in balance with the local source of surface water. When the basin is full, additional surface water flows down the streams. When some aquifers are first tapped, the pressure in many cases is enough to push the water to the surface, creating what are known as flowing artesian wells.

normal rainfall led to subsidence in some areas, and formation of a valley water conservation committee in 1920.

Nine years later, local voters approved formation of the Santa Clara Valley Water Conservation District to develop a plan to replenish the groundwater basin. Against the backdrop of the Great Depression, a $6 million bond measure on the local ballot failed in 1931, but residents approved a $2 million bond measure three years later.

Within a decade, the district had completed six dams of this so-called conservation project, an effort to conserve – and then use – local surface water that naturally flows into the ocean by recharging the groundwater basin.

The combined surface water system, recharge ponds and groundwater pumping, a program known as conjunctive use, has served the area well. ●

Groundwater in California

While the common image that comes to mind when one thinks of groundwater is a giant underground lake, groundwater usually is stored in the open spaces between rock, sand and other soil particles. These geologic formations are known as aquifers.

California's enormous groundwater aquifers hold ten times the amount of usable water that can be stored behind all its dams.

In the 1920s, the invention of the deep-well turbine pump and the electrification of rural areas put water hundreds of feet below the

W hen more water is extracted from a groundwater basin over a period of time than is recharged through natural – or artificial – methods, groundwater levels decline. In some basins, this decline can lead to land subsidence: the lowering or settling of the land surface.

The greatest amount of subsidence in California occurred in the San Joaquin Valley where, between 1925 and 1977, some 5,200 square miles of valley floor sank by at least one foot, up to 30 feet in some areas.

Geologists say the subsidence caused bridges and roads to crack and sink, changed on-farm irrigation grades and the slopes of natural streams and damaged more than 1,000 wells. The valley's subsidence problems eased in the 1960s and 1970s when surface water was first delivered through the Central Valley and State Water projects, and groundwater pumping was reduced.

The land surface near the community of Mendota suffered a nearly 30-foot drop between 1925 and 1977 because of groundwater pumping.

In this photo, Joseph F. Poland of the U.S. Geological Survey, who pioneered subsidence research in the San Joaquin Valley, stands near a benchmark of his studies. Signs on the power pole indicate the level of the land surface in 1925, 1955 and 1977. •

surface within reach for the first time. Farms flourished as hydro-power projects generated a source of affordable electricity. Acreage expanded and more land was put into production as irrigation supplies became more dependable. These new pumps also could deliver more water.

Conflict occurred as more and more wells were drilled to pump water from basins with declining water levels. With all overlying landowners having rights to the water

below their land and growth driving cities to seek an even greater share, disputes ended up in court. The first suit was filed in 1937 and some forty years later, most of the major groundwater basins in southern California had been divided and distributed by a court-appointed watermaster. Groundwater use in these basins is still closely monitored today. •

From Field to Florist's Shop

Pansies, poinsettias and potted plants are not what most people think of when it comes to crops. Yet in California, nursery products rank third in annual value on the state's top twenty agricultural commodities. If the cut flower category (which ranks ninth) were included, nursery products would rank second.

The state's thriving greenscape industry germinated with the Gold Rush as settlers brought with them vines, fruit trees and other plants from around the globe. These pioneers were enthralled with the ability to grow plants that previously had survived outdoors only in greenhouses, and such exotic species were in high demand. Miners who struck it rich had ready cash to spend when it came to landscaping their new estates.

The first nurseries opened in the late 1840s and within ten years, some forty such businesses had been established. The state's first recorded association of nurserymen formed in 1858.

One of the biggest greenscape success stories is the poinsettia. Today, no holiday table would be complete without this festive plant, known as the Christmas Eve flower. The Paul Ecke Ranch can be credited for

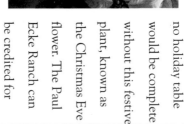

much of that. In the 1920s Albert Ecke and his son Paul began growing fields of the holiday flower and selling them at roadside stands in Hollywood and Beverly Hills.

The ranch moved to San Diego County in 1923 and the business grew as Paul Ecke traveled across the country teaching greenhouse growers how to cultivate the plant and market it as a holiday flower. •

Vegetable Variety

More than 300 varieties of crops and animal commodities are grown commercially in California. The state is the nation's leading producer of an alphabet of seventy-five diversified crops from apricots and asparagus to lemons and lettuce to wild rice and wine grapes.

And California is the only state in the United States that produces almonds, artichokes, dates, figs, kiwis, olives, persimmons,

pistachios, prunes, raisins, clovers and walnuts. Other agricultural commodities for which California is well-known include avocados, citrus, cotton, garlic, rice, safflower, spinach, strawberries and tomatoes.

The availability of regionally grown fresh fruits and vegetables spurred interest in the early 1990s in farmers' markets with consumers searching for organically grown produce.

Other futuristic trends that provide new niche markets for the state's farmers include ready-packed produce. Busy modern-day families increasingly favor the ability to purchase ready-to-go salads. Some experts believe that ultimately, one-fourth of the head lettuce grown in California will be used in these new salad-in-a-bag products. Other vegetables popular in such packs include broccoli, carrots, cauliflower and celery. Presliced fruit salads also are expected to increase in popularity. •

"The transformation of sagebrush wastes has become commonplace.
It is going on everywhere. But the wedding of the Colorado River
with the desert bearing its name, the birth of the Imperial Valley
and its miraculous growth, were not commonplace."

— Walter V. Woehlke —

"The Land of Before and After," Sunset magazine, April 1912

Southern California Agriculture

The roots of southern California agriculture date back to the Spanish missionaries. These early settlers were familiar with the Mediterranean climate of dry summers and mild winters, and established some of the first irrigated orchards and vineyards in Los Angeles.

Later settlers were awed by the frost-free climate of southern California. Here, farmers could raise crops virtually year-round, reaching those all-so-lucrative Midwestern and Eastern winter markets. All that was needed was water, and small and large irrigation works obliged.

Beginning in 1910, Los Angeles County was ranked the No. 1 agricultural county in the nation, a trend that continued until after World War II when the region began its farm-to-city transformation. Today the farm fields of Los Angeles County's San Fernando Valley as well as much of the farm belt found in Orange, Riverside, San Bernardino and

San Diego counties have been paved over by homes and shopping centers. Yet even with the switch from rural to urban living, high-value crops such as avocados, citrus, strawberries, cut flowers and nursery crops remain a vital part of the economy in these counties.

But it is the story of the Imperial Valley that represents the triumph of human ingenuity and technology over nature. Who could have imagined that the desert lands of this valley (once known as the Colorado Desert) would one day produce some $1 billion in crops each year?

George Chaffey and Charles Rockwood could. For it was under their direction that the first Colorado River water arrived in the region, renamed the Imperial Valley in 1901. The introduction of irrigation water set off a land boom and within eight months, 2,000 settlers had arrived and 100,000 acres were ready for cultivation. By 1905 the area's population had reached 14,000 and 120,000 acres of land were being farmed.

These farmers transformed what was once desert land of sage brush, cactus and mesquite into acres of vegetable fields.

Boosters created a saying:

"Egypt's Nile brings down the fruitful sediment of the highlands to spread it over a desert incapable of yielding even a spear of grass. The Colorado River is only required to furnish water for Imperial – she will do the rest." •

"… Here is a great plain, or rather a gentle slope, from the Pacific to the mountains. We are on this plain about twenty miles from the sea and fifteen from the mountains, a most lovely locality; all that is wanted naturally to make it paradise is water, more water. … The weather is soft and balmy – no winter, but a perpetual spring and summer. Such is Los Angeles …"

– William H. Brewer –
Up and Down California in 1860-64

"We might as well have had a catechism:

What is a farmer?

A farmer is a man who feeds the world.

What is a farmer's first duty?

To grow more food.

What is a farmer's second duty?

To buy more land.

What are the signs of a good farm?

Clean fields, neatly painted buildings,

breakfast at six, no debts, no standing water.

How will you know a good farmer when you meet him?

He will not ask you for any favors."

— Jane Smiley

A Thousand Acres

Tied to the Land

Whether backyard farmer with a summer vegetable garden or full-time grower with a field of tomato plants, there is something about the process of seeing a seed sprout, watching a plant grow, harvesting its fruit that stirs the soul. Even as we grow away from our agrarian roots, Californians remain tied to the land.

There is a commonly held, romanticized view of the family farm, a place where life is simpler and slower paced, with two or three generations working together to make a living off the land. Although this scene has historic roots – many of us have grandfathers or fathers who were farmers – it is a vision that often obscures the hard realities of farming. Agriculture, as an industry, is an endeavor of risk. Will there be enough water this year? Will there be too much? Will the weather wipe out a crop? Will the crop hold its value until harvest?

Farming in California is a combination of two seemingly juxtaposed realities: it is a way of life, but it is also a business. Farmers may like their independence, rural lifestyle and the connection to the land, but an agricultural operation must be profitable to support that lifestyle. And while many states were settled by small family farms, the agricultural history of California is one of large landholdings, beginning with the Mexican ranchos, continuing through the 1800s water rights battles to today's large corporate – albeit family corporations in many instances – farm holdings.

Since the Gold Rush, California's destiny has been one of growth. Lured first by dreams of striking it rich and later by the appeal of cheap land and inexpensive irrigation water, the rush to California was on. It continues unabated. Then as now, opportunity remains the watchword.

Some worry that farming will disappear in a state as devoted to growth as California is. They point to the transformation of Los Angeles County as an example. Others believe that while some individual farmers will go out of business, the industry as a whole will survive – and thrive – in California with higher value crops that serve to capture market returns through specialty crops, organic food and ready-packed produce such as salad-in-a-bag. ●

"This valley after the storms can be beautiful beyond the telling,
Though our cityfolk scorn it, cursing heat in the summer
and drabness in winter,
And flee it: Yosemite and the sea.
They seek splendor; who would touch them must stun them;
The nerve that is dying needs thunder to rouse it.

I in the vineyard, in green-time and dead-time, come to it dearly,
And take nature neither freaked nor amazing,
But the secret shining, the soft indeterminate wonder.
I watch it morning and noon, the unutterable sundowns;
And love as the leaf does the bough."

– William Everson

The Residual Years: Poems 1934-1948

Part of Our Everyday Lives

Water is the force that fuels the urban economy. Without water development, California would never have become the economic nation-state powerhouse it is. Yet the modern marvels of technology have made its presence – its very power – something we take for granted.

Consider a day in the life of California water. Its journey to the tap may begin in a river hundreds of miles away or beneath the ground just down the block. Cleansed of impurities at a nearby treatment plant, it flows through a maze of underground pipes to homes, offices and factories.

After the morning alarm rings, water is put to work. It flushes the toilet. It flows through the showerhead and the faucet, cleansing both body and teeth.

The force of its spray strips last night's dishes of the food it helped produce as the dishwasher hums and sings. In the laundry room, soap bubbles as water fills the washing machine.

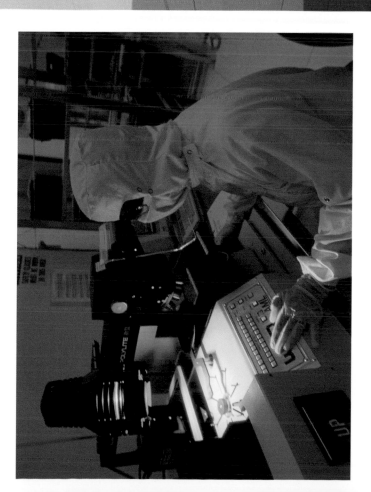

"The master condition not only of
any future developments in the West but of the
maintenance and safeguarding of what exists there now,
is the development and conservation of water production.
Water, which is rigidly limited by the geography
and climate, is incomparably more important than all other
natural resources in the West put together."

– Bernard De Voto –
Across the Wide Missouri

At the factory, it is water that cleans the soda bottles, produces the beer and processes the fruits and vegetables from the farm. It is an essential ingredient in the technological era: computer chip makers rely on millions of gallons of water to produce the chips that power the personal computer.

In the office building, it is water that brings us together to share our lives as we gather around the water cooler. The caffeine that jump-starts the business world comes only after water has met ground coffee beans.

With the turn of a valve, water rushes through yet another set of pipes to nourish the parks, school grounds, and golf courses that bring some open space to the concrete landscape.

At home in the evening, the rituals of washing the car, watering the garden or relaxing in the spa bring simple pleasures that touch the soul.

Water touches nearly every part of our everyday life. Yet most people don't give it a second thought when the water rushes from the tap. ●

he Folsom Powerhouse was built in 1895 to provide electricity to the young city of Sacramento, located 22 miles downstream. Electric power from the plant first reached Sacramento at 4 A.M. on July 13, 1895. Its arrival was greeted with much fanfare: a 100-gun salute courtesy of a unit of soldiers and a grand electric parade through the streets of downtown Sacramento.

Now a state historic landmark, the powerhouse ceased operation in 1952. Visitors can tour the facility located in the Folsom Lake State Recreation Area. Inside the power-house one can view the vintage generators, the massive General Electric transformers and the control switchboard, which is faced with Tennessee marble. ●

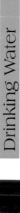

Drinking Water

Good quality drinking water does not just fall from the sky.

A tremendous amount of time and technology are employed to make water safe to drink. Between the water source and the tap is the treatment plant, where water is put through a variety of processes.

These technological advances are credited with virtually wiping out outbreaks of cholera and typhoid fever that occurred throughout the nation before the widespread use of chlorine as a disinfectant in 1914. •

Prior to municipal delivery systems, city residents got their water in a variety of ways. In San Francisco, barrels of water were delivered by water carriers; water routes similar to milk routes crisscrossed the city.

In Los Angeles, the earliest system consisted of the Zanja Madre or Mother Ditch. Men delivered each family's daily share in a barrel swung between the handles of a wheelbarrow. These barrels were later replaced with an ox-drawn cart, the water wagon. Eventually, a large water wheel with buckets attached to the paddles of the wheel, raised the water high enough so gravity could transport it to homes, fields and businesses through a system of hollowed logs. •

The Attraction of California

The first ads about the appeal of California's climate, resources and accessibility date back to perhaps as early as 1841. Fur trader James Ohio Pattie praised the region as "beautiful and sublime in scenery."

In the late 1860s, San Diego began selling its climate to health seekers.

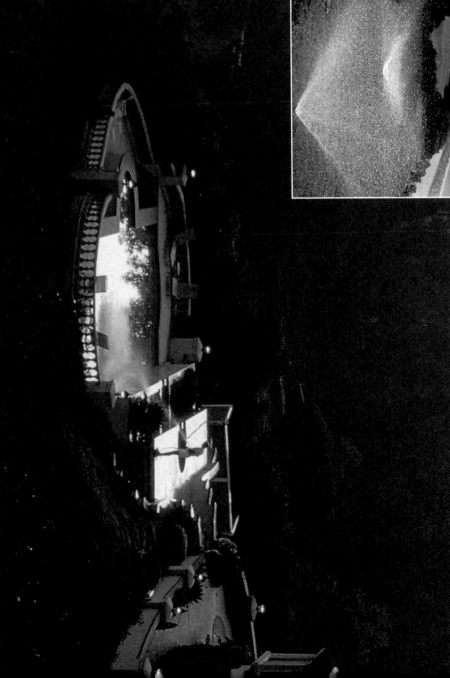

With completion of the Southern Pacific Railroad in 1870, others followed suit. People with tuberculosis, asthma and arthritis were encouraged to visit new health resort towns in Pasadena, Santa Monica and Palm Springs.

Shortly after World War II, California's population boomed. In his book *California: The Great Exception*, Carey McWilliams recounted the formation of one of southern California's first suburbs, Westchester.

"In 1940, Westchester was merely a name ... for a large, vacant area near the Los Angeles Municipal Airport.... In 1941 there were only seventeen widely scattered homes in the entire area; today 30,000 people live in Westchester. ... It is as though some- one waved a magic wand and a city had suddenly appeared." ●

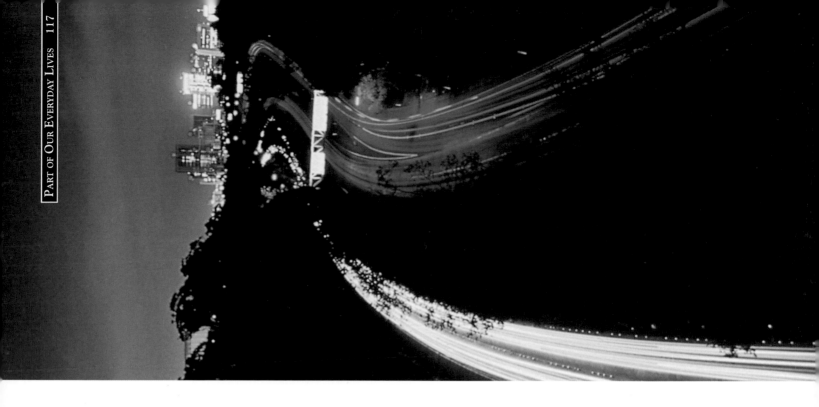

"Then out of the broken sun-rotted mountains of Arizona to the Colorado,

with green reeds on its banks, and that's the end of Arizona. There's California just over the river,

and a pretty town to start it. Needles, on the river. But the river is a stranger in this place.

Up from Needles and over a burned range, and there's the desert. And 66 goes on over the terrible desert,

where the distance shimmers and the black center mountains hang unbearably in the distance.

At last there's Barstow and more desert until at last the mountains rise up again, the good mountains,

and 66 winds through them. Then suddenly a pass, and below the beautiful valley, below orchards

and vineyards and little houses, and in the distance a city. And, oh, my God, it's over.

The people in flight streamed out on 66, sometimes a single car, sometimes a little caravan.

All day they rolled slowly along the road, and at night they stopped near water.

In the day ancient leaky radiators sent up columns of steam. . . .

People in flight along 66. And the concrete road shone like a mirror under the sun,

and in the distance the heat made it seem that there were pools of water in the road.

They drove through Tehachapi in the morning glow, and the sun came up behind them, and then . . .

suddenly they saw the great valley below them.

Al jammed on the brake and stopped in the middle of the road, and 'Jesus Christ! Look!' he said.

The vineyards, the orchards, the great flat valley, green and beautiful."

– John Steinbeck –
Grapes of Wrath

"Not in history
has a modern
imperial city been
so completely
destroyed.
San Francisco
is gone ...

Within an hour
after the earthquake
shock the smoke
of San Francisco's
burning was a
lurid tower visible
a hundred miles
away."

—Jack London
"The Fire"
Collier's

As early as 1882, San Francisco officials contemplated tapping the Tuolumne River in the distant Hetch Hetchy Valley as a source of water, applying for rights to divert the water in 1903 and 1905. The applications were denied. Then came the Great Earthquake and the Great Fire.

The quake that shook San Francisco awake at 5:13 A.M. on April 18, 1906, devastated the city. The fire that followed destroyed it. The water mains leading into the city broke during the temblor and the supply was quickly exhausted after fire started simultaneously in several different parts of the city. Firefighters turned to local streams, the Bay and even the sewers

in an effort to douse the flames. "Surrender was complete," Jack London wrote. "There was no water."

Two years later, the city's permit to dam and flood Hetch Hetchy Valley was approved by the Department of the Interior. Five years later, Congress passed the Raker Bill permitting the valley to be flooded. San Francisco residents, in turn, approved several bond measures to substantially improve and expand the city's local system of pipes and reservoirs, financing construction of an entire system of mains independent of the regular water system for fighting fires. ●

"How is it that water,
which is so very useful that life is impossible without it,
has such a low price — while diamonds,
which are quite unnecessarry, have such a high price?"

—Adam Smith —
The Wealth of Nations

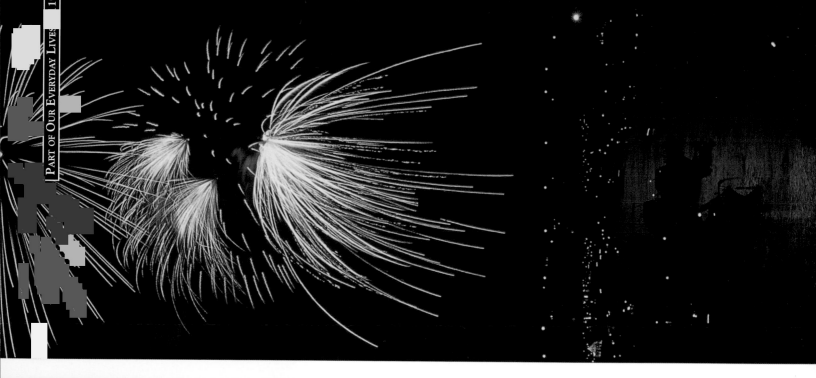

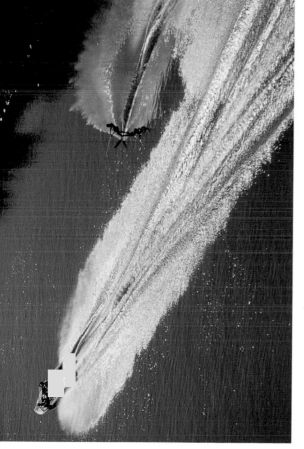

Soothing the Soul

Whether tranquil and serene or rushing white water, the sight of a river draws you to its edge.

The sound of its moving water relaxes your mind and soothes your soul.

The feel of cold water on a hot day is a welcome shock.

For our ancestors, the sight of water meant survival. At the end of a long overland crossing there was water for the animals, for themselves, for life.

Perhaps that old feeling is what draws us to live and work and play near water today.

As with many modern conveniences, the ease of water's arrival at the tap released us from the chore of fetching water and carrying it home.

With more time for leisure, the increasingly affluent society of post–World War II California sought the great outdoors to escape the congestion of growing urban areas.

Lakes, rivers and reservoirs became weekend destinations as residents took up boating, swimming, rafting and water-skiing. The state entered into a new era of managed recreation to prevent overuse and degradation. •

One of the most elaborate community pools was located in San Francisco. Opened in 1886, the Sutro Baths offered bathers one fresh water tank and six salt water tanks. The salt water tanks contained 1,685 million gallons of sea water, which filled and emptied at high and low tides.

Some 20,000 bathing suits and 40,000 towels were available for rent and bathers could play on slides, trapezes, swinging rings, springboards and a high dive. Galleries above the tanks were filled with seats for the many spectators who turned out each day. The three-acre site also included natural history exhibits, galleries of paintings and three restaurants. Admission was 25 cents for swimmers, 10 cents for spectators.

By 1937, attendance at the baths was down and the largest tank was converted to an ice-skating rink. Attendance continued to decline and the baths were abandoned. In 1966 the site was sold. As demolition work began — an apartment complex was planned for the site — a catastrophic fire broke out, destroying the eighty-year-old baths.

The fire and a subsequent movement to protect the site culminated in the baths' addition to the Golden Gate National Recreational Area. Today you can explore the remains of Sutro Baths and imagine the elegance of life here at the turn of the century. •

Once the ultimate extravagance for the rich, private pools have become a common feature for many middle-class families. Fly over any California suburb and it seems as if every other backyard contains a rectangle or circle of blue.

Both in climate and lifestyle the pool and its surrounding deck are part of the California home's regular living space. In many yards, the pool is the focal point in a landscape plan that is lush and elaborate.

As in other states, the first pools built in California were public pools. Here the community at large would gather to play. Early residential pools, known as estate pools, were a sign of status for the very wealthy. "For one to be able to walk directly from one's rooms ... to take a plunge in the crystal-like, invigorating water of one's own swimming pool is certainly to possess a rare, delightful luxury," *Sunset* reported in a 1926 issue profiling private pools in Beverly Hills.

By the 1940s, the modern gunite method of construction made pool ownership more affordable and widely available. While a semblance of the community pool remains at apartment complexes and private clubs, overall, the swimming pool has come to symbolize solitude rather than fellowship. As for a truly public facility, what has replaced the pool is the water park with its elaborate labyrinth of slides. •

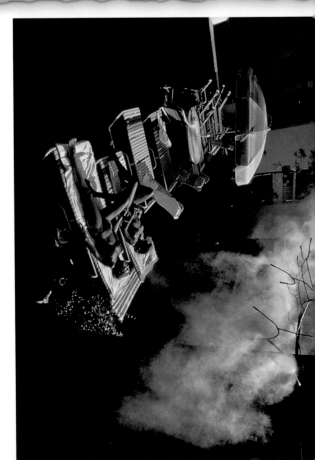

"Sometimes, when I can't see the pool,
I hear the non-stop suction of her swimming;
that aerated pummel, those thudding aspirates,
from which I have learned nothing at all, except to
admire such lightly worn virtuosity, and I know that
something perfect is being done, an ode to water is
writing itself longhand, long-leg. Sometimes her rhythm
slows, and I savor the lull between one bubble-beaded
thwack and the next, knowing that what I
hear is the perfect timing of float and push, of kinetic
energy almost gone and new energy supplied."

– *Paul West* –
Out of My Depths

outhern California's climate and the wide variety of fish off the coast of the Pacific Ocean served as an early tourist draw to Santa Catalina Island and the town of Avalon.

"It is the isle of summer in every sense," *Sunset* reported in its January 1901 issue. "The air is soft, like velvet on the cheek, and there is a crispness in the morning strangely at variance with the palms and bananas that top the neighboring knoll."

Guided boat tours offered visitors the chance to catch a record-sized black sea bass, yellow tail or tuna. •

CHRISS RINGSINS BOAT STAND.

"Once I said that trout fishing is a spiritual thing,

and after a lifetime, I know it is true. For that matter,

all fishing is a spiritual thing to a boy, no matter what he catches.

The sense of surprise, the eternal wonder of a fish coming out of the

water, the deep inherent sense of primitive accomplishment in getting

food by simple means, and the Pipes always playing softly

in the background – no wonder all boys love fishing, no wonder

all men, who really are boys at heart, feel the same."

– Sigurd F. Olsen –
Open Horizons

NEVADA

CALIFORNIA

Swimming at Tahoe

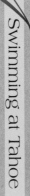

n the late 1800s, tourists began to visit Lake Tahoe in search of recreation or better health. To help their patients recover from tuberculosis, doctors in San Francisco often prescribed summer vacations at Lake Tahoe because the lake's fresh water and dry mountain air were believed to have restorative powers.

Bathers rarely strayed far from shore, however, because of tales that one would sink in the alpine lake's chilly, clear waters. Such talk related to early theories that the lake itself might be bottomless and observations that the bodies of drowning victims

rarely floated to the surface. The first person to successfully swim across Lake Tahoe was Myrtle Huddleston. On August 24, 1931,

"When the state of California becomes more populous, and the delicious summer climate of this elevated region, the exquisite beauty of the surrounding scenery, and the admirable facilities afforded for fishing and other aquatic sports become known, the shores of this mountain lake will be dotted with the cottages of those who are able to combine health with pleasure."

– John Le Conte –
Physical Studies of Lake Tahoe,
1884

she swam from Deadman's Point on the Nevada side to Tahoe City on the California side of the lake. It was a difficult swim as wind-swept waves caused her to veer off course, adding 7 extra miles to what was already a 13-mile effort. •

American River Park

In the days before air conditioning, Sacramentans spent their summers by the rivers. By the summer of 1931, a huge water slide on the American River at H Street attracted 3,000 bathers a day. •

"The river flows not past,
but through us, thrilling, tingling, vibrating every fiber
and cell of the substance of our bodies,
making them glide and sing."

– John Muir
John of the Mountains

To see a river cascading down granite cliffs, thundering through a cloud of mist is to partake in one of nature's most breathtaking and inspiring displays.

To behold a sparkling stream ducking in and out of forest shadows is one of nature's most poetic sights.

To watch a river meandering wide and slow through farm fields and past boat docks provides an image for one of our best daydreams – where would this river take me?

Utility and beauty best describe California's rivers and streams. For even as they cascade from mountains, tumble over rocks and course through canyons, their waters spill into reservoirs and swirl through penstocks generating water supply and power alike to fuel the state's economy. The rivers are the state's lifeblood and recreational playgrounds. Boats ride the waves while fishermen fish from

the shore. Rafters challenge the force of a river's flow while swimmers seek its cool relief.

Geography, climate and human decisions created today's waterscape. Upstream dams control flooding, store water for later use and allow for more consistent year-round flows; some rivers, even in the north part of the state, would run dry during summer months before dams. In the southern end of California, river flow is much more unpredictable. Many streams are ephemeral, reappearing only with rainfall. Sudden, intense rainstorms can cause these streams to appear and become torrential rivers, catching the unaware off-guard, even sweeping them to their death. ●

"The river turns on itself,
The tree retreats into its own shadow.
I feel a weightless change, a moving forward
As of water quickening before a narrowing channel
When banks converge, and the wide river whitens;
Or when two rivers combine, the blue glacial torrent
And the yellowish-green from the mountainy upland, –
At first a swift rippling between rocks,
Then a long running over flat stones
Before descending to the alluvial plain,
To the clay banks, and the wild grapes hanging
from the elm trees."

– Theodore Roethke –
The Collected Poems of Theodore Roethke

A Shift in Philosophy

Mark Twain's disparaging observations about Mono Lake did not portend the singular place this isolated, inland, saline body of water would come to hold in the California water story.

For it was here at this Eastern Sierra lake (right) that a 1970s movement to "Save Mono Lake" led to a landmark 1983 court decision, marking a turning point not only for Mono Lake itself, but for the environmental movement as a whole.

In its 6-1 ruling, the California Supreme Court determined that the public trust doctrine applied to Mono Lake and its environmental values. The doctrine, rooted in Roman Law and common law, requires that the sovereign or state hold designated resources in trust for the public.

Although it would be another ten years before the city of Los Angeles would be forced to reduce its dependence on the tributaries feeding Mono Lake and the lake would begin to rise, the court's decision gave the environmental movement immediate legal and political clout.

The movement itself began in the late 1960s and early 1970s as everyday citizens' concerns about water pollution, free-flowing rivers and disappearing plants and animals gave rise to powerful laws designed to protect the nation's natural resources.

The shift in philosophy from one of developing resources to protecting resources caught California's traditional water users by surprise. Conflict soon followed as the different interest groups faced off over additional dams, Delta diversions and water rights. Two decades of lawsuits and bitter debate would follow, with each side having its share of victories and defeats. ●

"Mono Lake lies in a lifeless, treeless, hideous desert . . .
This solemn, silent, sailless sea — this lonely tenant of the loneliest spot
on earth — is little graced with the picturesque. It is an unpretending
expanse of grayish water, about a hundred miles in circumference with
two islands in its center, mere upheavals . . . of the dead volcano,
whose crater the lake has seized upon and occupied."

Decades before the advent of the modern-day environmental movement, John Muir served as president of the newly incorporated Sierra Club. Founded in 1892, the organization was an early advocate of protecting and preserving wilderness lands.

Part of the club's initial mission was "to enlist the support and cooperation of the people and government in preserving the forests and other natural features of the Sierra Nevada." Early victories included defeating a plan to reduce the boundaries of Yosemite and establishing the Devils Postpile National Monument.

As the nation's oldest environmental organization, the Sierra Club remains a vital force in the ongoing effort to protect natural resources by lobbying against dams and for laws mandating clean air and clean water. •

"Mono's beauty, power, and worth comes from more than birds, shrimp, tufa, islands or people alone. It comes from all these things together. It comes from wholeness."

— *David Gaines* —
Mono Lake Committee

Mark DuBois guiding rafters downriver.

"The Stanislaus brought out feelings of reverence for the natural world, where rivers are a source of life. . . . The river was a sacred and holy place."

– Tim Palmer
Endangered Rivers and The Conservation Movement

A Turning Point

It was perhaps the most dramatic moment in the long fight over New Melones Dam. Rafter and environmentalist Mark DuBois chained himself to a rock in the Stanislaus River Canyon to protest plans to fill the reservoir, the last of the big on-stream dams built in California.

DuBois' actions attracted national attention to the controversy. Part of the Central Valley Project, New Melones was authorized in 1944. Construction began in 1966, galvanizing the state's river-preservation movement. At issue was the flooding of a popular stretch of white-water rapids and the West Coast's deepest limestone canyon.

As proposed, the dam's 23-mile-long reservoir would inundate such rapids as Cadillac Charley, Widow Maker and Devil's Staircase. A campaign for a smaller reservoir that would preserve the rapids failed. Dam opponents continued the fight even after the dam was finished in 1978.

Plans to raise the water level to test the dam's turbines prompted DuBois' action. In a letter to the U.S. Army Corps of Engineers, he said as of May 21, 1979, he would be "permanently anchored to a rock" in the canyon so his body would be at least partially immersed if they raised the level.

By the time DuBois voluntarily released himself a week later, Gov. Edmund G. "Jerry" Brown had appealed the issue all the way to the Carter White House. DuBois declared victory after the Corps agreed not to flood the rapids in the immediate future. "The era of dam building is over," he told *The Sacramento Bee.* "I feel like this is a major turning point – the Corps turned back."

The reservoir's ultimate level, however, was to be determined by the Bureau of Reclamation. The battle over the facility moved to Congress where opponents pursued federal Wild and Scenic Rivers designation for the Stanislaus River. The proposal fell short by one vote in a key committee. Voters rejected a subsequent ballot measure to grant the river state protection.

In a last-ditch attempt to have Carter block the dam, five members of the Stanislaus Wilderness Committee chained themselves to a rock in 1981, sending the key to Carter. This effort, too, failed and the reservoir filled during a high runoff year in 1982-1983. •

Valley Project. By the 1968 ground-breaking ceremony, however, conservation groups had launched a major effort to block New Melones Dam and Congress had passed the Wild and Scenic Rivers Act.

In 1972, three environmental groups sued the federal government to block construction of Auburn Dam. As the court battle raged on, an earthquake occurred near the town of Oroville. The 1975 quake was some 40 miles away, but was on the same fault line – generating concern about Auburn Dam's seismic safety.

New seismic considerations were built into the design, but changes in federal-state cost sharing, as well as growing opposition to Auburn from environmental groups and national taxpayer organizations over long-term impacts and costs delayed the dam. As controversy continued through the 1980s and 1990s over whether to complete the project to boost Sacramento's flood protection, environmental groups remained opposed to Auburn Dam. •

Earthquakes, the environmental movement and economics stalled completion of Auburn Dam. Planned for the North Fork of the American River, the dam was authorized by Congress a year after the 1964 Christmas flood threatened to inundate Sacramento. Folsom Dam saved the city, but officials calculated one more day of rain and record runoff would have overwhelmed the dam.

The environmental movement hadn't really begun when Auburn Dam was authorized as part of the Central

Scenic Rivers and Decisions

1998 marked the thirtieth anniversary of one of the most powerful environmental laws, the national Wild and Scenic Rivers Act. For some, the act remains the tool of choice to stop the construction of new dams and diversion structures. Most of the protected rivers in California had been identified at some point for a potential project, varying from small hydropower dams to large reservoirs.

Signed into law by President Lyndon Johnson in 1968, the national Wild and Scenic Rivers Act was designed to balance interests in water development and environmental protection by protecting sections of rivers from further water development. In 1972, California's legislature passed its own state act to protect the north coast rivers, including the undammed Smith River.

> *"Why save a river?*
> *Because it is the*
> *unspoiled Eden,*
> *right here, available*
> *to all. Why save a*
> *river? Because a*
> *river is a river."*
>
> **– Tim Palmer**
> **Endangered Rivers**
> **and The Conservation**
> **Movement**

Tale of the Trinity

The tale of the Trinity River is a classic confrontation between fishermen and farmers. In its natural state, the Trinity flowed east to west. The river once supported large runs of salmon and steelhead as it flowed into the Klamath River and on to the Pacific Ocean. With completion of the Trinity River Diversion in 1963, up to 90 percent of the river's flow was diverted east through a tunnel into the Sacramento River where it augmented Central Valley Project supplies. The Trinity's waters helped irrigate thousands of acres in the Central Valley. But even as the river helped the valley bloom, its once-plentiful fish populations began to decline.

More than thirty years of controversy followed as Trinity County residents fought to increase releases from the dam to boost fish runs. Farmers resisted, fighting to maintain their water supplies. As in other parts of California, the commercial salmon fishery added fuel to the fire as water users contended over fishing caused the salmon's decline.

Today, some 340,000 acre-feet flows down the Trinity compared to 120,000 acre-feet after the dam was first built. But as much as 1 million acre-feet more is diverted into the Sacramento River. In early 2000, both sides were awaiting word whether releases would be boosted to approximately 50 percent of the river's natural flow. ●

River stretches may be designated as "wild," undeveloped and accessible only by trail; "scenic," essentially wild but occasionally accessible by road; or "recreational," accessible by road with some development allowed. All classifications prohibit new dams and major diversions. California has over 1,900 river miles under federal and/or state protection. ●

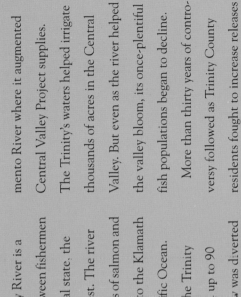

Although it failed to gain Wild and Scenic status for the Stanislaus, the conservationist organization Friends of the River – formed to fight New Melones Dam – was key to victories preserving other rivers including, the King, Kern, Merced, Tuolumne, Klamath, West Walker, East Carson, Sisquoc and Big Sur rivers and Sespe Creek. Other protected rivers under either the state or federal acts are the entire Smith River and sections of the Trinity, Van Duzen, Scott, Eel, Salmon, Feather and American rivers.

Peripheral Canal

In 1960, Gov. Edmund G. "Pat" Brown's "one state" campaign successfully united northern and southern California voters and secured a narrow victory of the bond to finance the State Water Project (SWP). California's north-south division proved much more contentious twenty-two years later during his son's administration.

At issue was construction of an SWP Delta transfer facility, the Peripheral Canal. As designed, the 43-mile, dirt-lined canal would take water directly from the Sacramento River and transport it around the Sacramento-San Joaquin Delta to the export pumps for shipment south via the California Aqueduct.

Within the water community, controversy over the canal had raged since it was identified as the transfer facility of choice in the mid-60s. When Gov. Edmund G. "Jerry" Brown took office in 1975, he ordered a reappraisal of the project. Two years of hearings followed in which the opposing factions hardened their positions. Brown turned to the legislature, instructing members to produce a bill authorizing the canal.

His father's strategy had been to provide something for every region of the state in his water package. Brown tried the same approach, linking constitutional protection of the North Coast rivers from further development and Delta water quality in exchange for the canal and supporting facilities. In some ways the 1980 package was an early attempt at balancing environmental protection with continued water development. But no one liked the proposal. Environmentalists deemed the protections too weak; water users – especially the farm community – saw them as too strong.

The Peripheral Canal bill package, known as SB 200, passed the Senate in January 1980. The Assembly followed suit in early July and Brown signed the bill into law on July 18. Meanwhile,

"Powerful bureaucracies and special corporate interests have played monumental roles in shaping modern California. The fundamental explanation for the contemporary waterscape, however, is to be found in the fact that the public has historically tolerated, and until recently, vigorously supported great water projects. Individual citizens cast votes for the Owens River aqueduct, ... the Hetch Hetchy project, the Colorado River aqueduct, the original Central Valley Project and the State Water Project."

– *Norris Hundley –*
The Great Thirst

the legislature voted to put the constitutional environmental protection changes on the November 1980 ballot. Voters approved Proposition 8 by a margin of 54 to 46 percent.

But there was a catch – the environmental protection measures would go into effect only if the canal were built, and opponents were already at work gathering signatures to force a voter referendum on SB 200. They succeeded and the referendum on what was now known as Proposition 9 was slated for the June 1982 ballot.

The ensuing fight over the canal became an epic political battle and one of the most expensive campaigns to date – the combined total spent by both sides exceeded $6 million.

The campaign clearly illustrated the regional dispute over water as northern Californians fought to keep the south from "stealing our water." It also highlighted the growing water development environmental protection split, which created strange bedfellows

as environmentalists joined forces with some farm interests to defeat the measure.

Opponents focused their efforts on the cost of the project. Their slogan "It's just too expensive" appeared on billboards throughout the state.

Memories of the 1976-1977 drought had faded some by 1982, which may explain why proponents' message that "We need the water" failed to garner much support.

On election day, voters in northern California rejected the project by a

9-1 margin. Only eight of the state's fifty-eight counties – Kern and the seven counties south of the Tehachapi mountains – voted in favor of the June 1982 ballot measure authorizing the canal's construction.

It was a major victory for the environmental movement and the first defeat of a major water project in some sixty years. •

root zone. But this practice requires good soil drainage.

To improve drainage, farmers install pipelines underground to collect the salty water. Where to dispose of this water is a controversial issue – especially on the west side of the San Joaquin Valley. At one time, a large drain was planned to carry this water away. But the drain was never completed, and the water collected at a place called Kesterson Reservoir.

In 1983, scientists discovered hundreds of dying adult birds and deformed or stillborn embryos at Kesterson, which was jointly managed to provide habitat for waterfowl. Studies revealed that high concentrations of selenium, a trace element that occurs naturally in some soils, had reached toxic levels in the birds. The drainage flow was halted in 1986 and the contaminated ponds filled in. However, disposal of drainage water remains a serious problem for this region. ●

ne of civilization's oldest problems is the buildup of salts in soil through farming. In the Middle East, what was once the fertile crescent is today mostly desert, partly because of salt buildup. In California, where irrigated agriculture is high-tech, farmers apply extra water to wash salt from a plant's

Silence Turns to Action

The first embers of the modern-day environmental movement were sparked by Rachel Carson's 1962 book *Silent Spring*.

The controversial book documented the dangers DDT and other pesticides posed to the natural world. *Silent*

Spring shocked a country that had grown accustomed to the advantages of post-World War II technology; it was the first look at the downside of such advancements.

After twenty years of a post-World War II boom — especially in California

"No one has the right to use America's rivers and America's waterways, that belong to all the people, like a sewer. The banks of a river may belong to one man or one industry or one state, but the waters which flow between the banks should belong to all of the people."

– President Lyndon Johnson signing the Clean Water Act of 1965

— Americans looked around to discover that their rivers and beaches were fouled with industrial waste and sewage.

When Ohio's heavily polluted Cuyahoga River caught fire, the image flashed around the world and served to dramatically illustrate the movement's cause.

Government responded. Led in large part by Interior Secretary Stewart Udall during the Kennedy and Johnson administrations, Congress passed a series of environmental laws: the federal Clean Air Act of 1963, the Wilderness Act of 1964, the Wild and Scenic Rivers Act of 1968 and the National Environmental Policy Act of 1969.

Earth Day in April 1970 showed the strength of this grassroots movement as rallies and demonstrations were held at thousands of college, high school and even elementary school campuses in every state.

Subsequent laws passed during the Nixon, Ford and Carter administrations served to further strengthen the environmental movement: the Clean Water Act of 1972, the Endangered Species Act of 1973 and the Federal Land Policy and Management Act of 1976.

With the Clean Water Act, the federal government established as a national, shared priority ending the discharge of pollutants into water. Virtually every city was required to build a sewage treatment plant; nearly every factory was forced to devise a means to clean up its discharge pipe.

It was the people versus them: cities and factories. The Clean Water Act has worked well at making much of the nation's water safe for fishing and swimming. But problems remain today with nonpoint source pollution.

It is a more difficult challenge to clean up diverse, multiple points of pollution because it becomes the people versus themselves. They are the polluters and the solution rests in a host of changes – major and minor – to everyday activities. ●

Clean-up and Prevention

Thousands of rural Californians safely drink well water just as it comes out of the ground. Groundwater in other areas requires only minimal treatment before it can be used.

But some regions of the state are dealing with extensive contamination, resulting in lengthy, expensive clean-up programs.

Some of the problems stem from practices once thought to be benign. It was once commonly believed, for example, that the soil would filter out many chemicals used above ground before they reached the groundwater aquifer far below.

Beginning in the 1970s, routine testing of groundwater revealed the hard truth of such practices as water suppliers discovered pesticide contamination from farming operations, hydrocarbons and solvents from defense and computer plants, and benzene from leaking underground gas tanks.

While removal of such pollutants remains a high priority, the No. 1 groundwater mantra among water officials today is "prevention, prevention, prevention" in an effort to guard against further contamination. ●

A Resource is Tested

The salmon's life cycle inspires wonder. Tiny fry emerge from eggs in the cool freshwater of a river, wind their way downstream and adapt to the salt water conditions of the ocean.

Three to four years later, the adult salmon – weighing as much as fifty pounds – respond to instinct and return from the ocean. They struggle to swim perhaps hundreds of miles upstream to spawn and then die in the same river in which they were born.

Salmon have made the river-to-sea to-river journey for thousands of years. It is an ancient rhythm. Even in this modern era, young and old alike visit fish hatcheries or gather along river banks to watch the battered, trium-phant fish return to the place of their birth.

The number of salmon historically ebbed and flowed with the state's fluctuating water supply and other natural factors. Yet it was the precipi-

No one knows exactly how many chinook salmon once made the yearly journey through the Delta upstream to spawn in the Central Valley's great rivers. Colorful anecdotes – "There were so many salmon you could walk across their backs" — and canning records indicate that millions once migrated between the Pacific Ocean and the headwaters of the Pit, McCloud, Sacramento, Feather, American and San Joaquin rivers. In 1882, the commercial salmon catch from the Sacramento River alone was a record 12 million pounds.

From 1873 to 1910 as many as twenty-one canneries in California annually processed 5 million pounds of salmon from the Sacramento and San Joaquin rivers and their tributaries. In those days salmon fishermen turned to the rivers to catch the adult fish as they returned to their home stream to spawn. To bolster the commercial fishing industry, the Baird Fish Hatchery was established in 1872 on the McCloud River, the first artificial propagation of salmon in California. Despite it and other early hatcheries (Battle Creek in 1897 and Mill Creek in 1902) overfishing, along with mining debris, dams and diversions, acid drainage from hard rock mining and the construction of the railroad caused a drastic decline in salmon populations. •

tous drop of the Sacramento River's winter-run chinook during the severe drought of the late 1980s that generated the most concern.

In the view of some, the winter-run was headed toward extinction. In stepped the Endangered Species Act, a powerful law designed to protect species from the fate of so many others, such as the now-extinct California grizzly bear.

The winter-run's designation as an endangered species has forced a change in traditional water project operations. Where decisions were once made solely on water and irrigation needs, operators of dams and diversions both large and small must now consider the winter-run before storing, releasing and pumping water. •

"To waste, to destroy, our natural resources, to skin and exhaust the land instead of using it so as to increase its usefulness, will result in undermining in the days of our children the very prosperity which we ought by right to hand down to them amplified and developed."

– Theodore Roosevelt –

Bitter Divisions

Played out against the backdrop of the 1987-1992 drought and requirements to protect the endangered winter-run chinook salmon, the intense political debate over the Central Valley Project Improvement Act (CVPIA) illustrated the bitter divisions between environmentalists and farmers.

Signed into law by President George Bush just days before the 1992 general election, the landmark CVPIA brought sweeping change to the federal water project. The act allocated more water to fish restoration and wetlands enhancement, in effect making the environment a CVP contractor.

With passage of the act, agricultural users – for the first time – would be permitted to sell their water to a city outside the CVP service area.

Environmentalists were triumphant. "The environment is now a contractor of the CVP," one said. Urban purveyors trumpeted the new law as "tanta-

mount to creating a new reservoir overnight."

For the agricultural users of the CVP, the law was a major defeat. It demonstrated the power of the alliance between conservationists and municipal water providers in what is the three-legged stool of California water issues – agricultural, urban and environmental interests.

The outcome reflected the shift of political clout from farms to cities not only in California but throughout the West. It was the decision by congressional advocates of the CVPIA to link it to a dozen other water bills that aided its passage.

Although agricultural water use still dominates in California and other Western states, population equals representation in Congress and the region's growing urbanization has eroded the rural areas' traditional power.

The bill split the traditional ag-urban caucus as the other states

"And when the
dawn-wind stirs
through the ancient
cottonwoods,
and the gray light
steals down from
the hills over the old
river sliding softly
past its wide brown
sandbars — what if
there be no more
goose music?"

– Aldo Leopold
A Sand County
Almanac

Politics, Biology and Engineering

Often characterized as a case of David versus Goliath, the 3-inch Delta smelt's listing as a threatened species brought big changes to the large export pumps in the south Delta.

The Central Valley and State Water projects were designed to capture high flows during wet periods. Historically, February and March were the main months for exporting water from the Delta. In the early '90s, exports were modified these months to protect endangered winter-run chinook salmon as they migrate through the Delta – pushing the export window to the spring or later.

Yet this is when young Delta smelt are found in the central Delta. Adult smelt move upstream from Suisun Bay in February, spawning in April and May. Water exports are closely monitored, and modified as necessary, to prevent smelt from being drawn south into the pumps and killed. Generally, young smelt move away from the pumps by June, easing restrictions. In 1999, however, they lingered near the pumps into early July and as reservoir levels south of the Delta dropped, conflict flared.

The effort to protect Delta smelt, which has only about a one-year life span, illustrates how closely each action is linked and how difficult it is to maintain balance in water system management. Seventy-five percent of the water originates in northern California; yet 80 percent of the demand is south of the Delta. With the need to protect fish, the Delta export window has been moved to the summer and fall – when the main source of water is from reservoirs, not storm runoff.

Fear that it will be a dry year and that the projects will be unable to fill reservoirs south of the Delta increases tension when water project's must modify operations to protect endangered fish. For better or worse, most of the state's residents and thousands of farms depend on water from the Delta. In the end, the system is driven as much by politics as biology and hydrology. •

found that the only way to secure their own projects was to support the CVPIA.

In California, the fight over the CVPIA was a major milestone of the environmental movement. Through use of the judicial, legislative and political arenas, conservationists had secured victory in the Mono Lake fight and the Peripheral Canal battle. They had achieved Wild and Scenic status for many river stretches and endangered species status for many declining species.

By the mid-1990s, environmental groups had secured their place at the negotiations table dealing with the future of water management in California.

However, the fight over the CVPIA did not end with the bill's passage. Conflict has continued over the law's implementation even as the three interest groups seek to find a balance on a host of economic and environmental issues. •

"A river is a body of moving water large enough to
occupy one's mind ... Rivers are metaphor for change,
dreams of history. They move by, tangling the threads of time,
braiding their waters into mixtures of the moment,
open systems hinting at a single destiny,
a predetermined and overwhelming need for the sea."

– Lyall Watson –
The Water Planet

"In the 'recreation versus civilization' controversy, whatever stand a man takes, he is discovered sitting opposite himself at the witness table. Does he rank the need for places to fish and get away from civilization ahead of the needs of civilization itself? The pressures of industrial society sometimes seem irresistible. … What is the farmer to do who is also an avid angler? What is a fisherman to do who enjoys electric conveniences at home as much as the next man? He will be torn between his two selves until he recognizes that the two selves are one. Before he can win the battle in the hearing room he must win the battle with himself: when people become so truly civilized that they are able and willing to control their numbers — only then will they stop encroaching upon their heritage."

— Erwin Cooper —
Aqueduct Empire

The Balance

The first century after California gained statehood was defined by the effort to develop cities, farms and industries. Water helped fuel that growth and the technological ability to move it, use it and redirect its path provided the state's citizens with prosperity and protected them from floods.

Yet these programs came with an environmental price – the alteration of the state's natural waterscape dried up wetlands, left some streams high and dry and blocked migrating fish from reaching historic spawning grounds. It was in the 1960s that the tide began to turn as conservationists demanded that some water be put back to benefit nature.

Today, the California water story has entered a new chapter. It is a chapter devoted to better balance between the environment and the state's economic engine. The chapter is still being written as people try new ideas, new alliances, new thinking.

Balance is difficult in a system built on adversarial processes. Competition for water began with the rush to find gold and continues in an era of increasing demand and decreasing supply. Conflict remains. But at the same time there is more interest in

"I believe that eventually,

perhaps within a generation or two,

they will work out some sort of compromise

between what must be done to earn a living

and what must be done to restore health

to the earth, air and water."

– Wallace Stegner –

Where the Bluebird Sings to the Lemonade Springs

seeking consensus on different issues and cooperating on various programs.

This new chapter is being shaped by a variety of forces – the power of drought, the power of the political process, the power of environmental protection laws. Yet it is something else that pushes us toward this new system of environmental protection and economic productivity – it is the power of water to provide not only goods and services, but peace of mind. Water offers an escape. The roar of rushing water draws us to a river's edge. The serene reflection in the lake soothes our soul. We as a society want

to have clean rivers to swim in, fish to catch, birds and animals to watch. We want to ensure that our children will enjoy the natural world even as we reap the benefits of the past generation's efforts to develop California.

It is for these reasons that everyone from competing water users to the neighborhood watershed group has joined forces to try to resolve a host of environmental issues and restore wetlands, fish populations and the neighborhood creek. Such efforts restore our faith in ourselves. •

"Conservation is a state of harmony between men and land. By land is meant all of the things on, over, or in the earth. Harmony with land is like harmony with a friend; you cannot cherish his right hand and chop off his left. That is to say, you cannot love game and hate predators; you cannot conserve the waters and waste the ranges; you cannot build the forest and mine the farm. The land is one organism. Its parts, like our own parts, compete with each other and co-operate with each other. The competitions are as much a part of the inner workings as the co-operations. You can regulate them — cautiously — but not abolish them."

– **Aldo Leopold** –
A Sand County Almanac

Consensus in a Political Quagmire

The defeat of the Peripheral Canal did not end the dilemma over the Delta's dual roles: switchyard for the state's biggest water projects and home

Interior Secretary Bruce Babbitt, left, and California Gov. Pete Wilson announcing the 1994 Bay-Delta Accord.

to a rich and diverse ecosystem of plants and animals. From questions about water transfers to water quality, endangered species protection to growth, the Delta debate continues to dominate California water politics.

For politicians, the canal campaign clearly illustrated one major point: the Delta was a political quagmire.

The point was brought home to Jerry Brown's successor, George Deukmejian, when a smaller water transfer facility – dubbed Duke's Ditch – went down to decisive defeat.

Adding to the difficulty were the shifting alliances among the three main stakeholder groups, environ-mental, urban and agricultural. None had sufficient power for complete victory, yet each had enough pull with voters and lawmakers to thwart the plans of the others. Sometimes, urban and ag groups would join forces; other times, urban and environmental groups would become allies on an issue.

As it became clear that no one interest group could prevail, a move-ment toward consensus gradually emerged. With the 1987-1992 drought providing further impetus, leaders of the three main groups began meeting behind closed doors with a facilitator on a regular basis in 1990. It would be another four years, however, before months of meetings and hours of negotiations would end in a five-year peace with the signing of the so-called Bay-Delta Accord of 1994.

The accord was heralded on all sides as a landmark agreement. Here, it finally seemed, the environment and economics would be balanced. The accord increased the amount of water dedicated to preserving Delta habitat while promising to restrict additional demands if more species were added to the endangered list. For water users, this agreement would provide them with more certainty in annual supplies. The accord also brought about a major shift in state-federal relations as various agencies – with often compet-ing missions – pledged to end their conflict and cooperate. •

A Cooperative Spirit

California is familiar with the political tension than can exist between the federal government and the Western states when it comes to water.

Water rights – the right to use certain supplies – are governed by state law. However, the federal government has the authority to regulate water quality and its enforcement of the Clean Water Act can affect a state's ability to allocate water.

In California, strained relations with the federal government over Bay-Delta water quality standards were exacerbated by controversy over measures taken to comply with federal environmental mandates in the late 1980s and early 1990s.

One of the major breakthroughs that helped lead to the 1994 Bay-Delta Accord was a new cooperative spirit between the state of California and the federal government on Delta

issues. Coordination among top policy makers from both the state and federal water and environmental agencies helped forge an agreement on water quality and endangered species issues.

With the accord, the new state-federal effort was formalized into a collaborative process known as CalFed.

Since 1995, the agencies that comprise CalFed, along with input from the stakeholder groups, have worked to develop a long-term solution to the Bay-Delta dilemma. With implementation of a collaborative plan the ultimate goal, CalFed's job has not been easy.

Yet just the fact that water project engineers and fishery biologists from state and federal agencies are striving to find a solution remains a positive development. •

Wetlands Restoration

Wetlands were once valued only for the land beneath the water. As the state's cities grew and farms expanded, thousands and thousands of acres of wetlands were drained of water, their soil converted to another use. There was little early comprehension of how draining a wetland changed the entire water system.

Today, wetlands are valued for a wide variety of reasons: for providing fish with spawning and rearing habitat; for detaining and slowing flood flows, reducing the size of floods; for absorbing and filtering pollutants that otherwise degrade streams and groundwater aquifers; and for providing places for groundwater recharge.

Where the government once provided financial incentives to private landowners who drained and reclaimed wetland areas, the new projects of the 1990s encouraged private landowners to restore wetlands.

Throughout the country, efforts are underway not only to protect what

wetlands remain, but to expand wetland acreage. California officials estimate that since 1993, nearly 200,000 acres of wetlands have either been enhanced or acquired and protected. Such actions are aimed primarily at protecting and increasing the populations of waterfowl – California provides critical wintering habitat for millions of waterfowl migrating along the Pacific Flyway between Canada and Mexico.

These actions are guided not only by laws and regulations, but by a growing recognition of how the natural landscape can allow us to live our modern lives.

If wetlands help reduce floods by slowing flows, they can reduce property damage and loss of life. If wetlands filter out pollutants, they can help us clean up our polluted rivers and lakes. If wetlands help recharge groundwater levels, they can help us alleviate overdraft. ●

"For millennia, water lay over the land.

Untold generations of water plants, birds, animals, insects, lived, shed bits of themselves, and died. I used to like to imagine how it all rifted down, lazily, in the warm, soupy water – leaves, seeds, feathers, scales, flesh, bones, petals, pollen – then mixed in the saturated soil below and became, itself, soil. I used to like to imagine the millions of birds darkening the sunset, settling the sloughs for a night ... the riot of their cries and chirps, the rushing hough-shh of twice millions of wings, the swish of their twiglike legs or paddling feet in the water, sounds barely audible until amplified by the millions. And the sloughs would be teeming with fish: shiners, suckers, pumpkinseeds, sunfish, minnows, nothing special, but millions or billions of them."

– Jane Smiley –
A Thousand Acres

The New Projects

Much of the impetus to develop the state's water at the turn of the century stemmed from its seemingly unstoppable growth. As Kevin Starr wrote in his book *Americans and the California Dream, 1850-1915,* "From the beginning, California promised much. While yet barely a name on the map, it entered American awareness as a symbol of renewal. It was the final frontier: of geography and of expectation."

Today, the state still faces pressures of growth as people continue the search for that all-so-elusive California Dream. Combined with the recent mini baby boom, officials estimate that California's population will increase from 32.1 million to 47.5 million in the next two decades.

Just as they did a hundred years ago, the state's urban leaders are expanding water supplies to meet this increased demand. Unlike the projects of old, however, these new projects do not focus on damming rivers and streams.

The new focus is on stretching existing supplies, transforming dry valleys into water storage reservoirs or saving water underground.

In short, these new projects seek to provide benefits for humans with the least harm to nature.

In Contra Costa County, the local water district built a new reservoir in the hills east of San Francisco. At Los Vaqueros Reservoir, water is pumped from the Delta during high flows each spring and stored for use later in the year, when Delta water grows too salty. In Kern County, high flows from the Kern River are spread out over 20,000 acres of land, slowly moving through the soil to recharge the groundwater aquifer below.

Along with off-stream and underground storage, water conservation, water recycling and water marketing are the projects of the future. Desalination of sea water — an expensive source today — may become more prevalent as new technologies emerge. ●

t was a ceremony with great fanfare. A mile-long dynamite blast and red, white and blue smoke bombs heralded the groundbreaking of a new southern California reservoir. The celebration had echoes of the past, but the year was 1995 and this modern-day reservoir is nothing like the projects of old.

To start, the Metropolitan Water District of Southern California's new Diamond Valley Lake is an off-stream facility; no river was dammed in order to create this artificial lake.

In addition, the reservoir site in Riverside County is surrounded by some 13,000 acres of land set aside as a preserve for the endangered Stephen's kangaroo rat and several other species of wildlife. ●

A New Water Ethic

Once the rallying cry of droughts, "save water" has become an every-day goal of water suppliers through-out California.

Many conservation measures initiated as emergency drought measures have become institutional-ized through state law and regula-tions. Low-flush toilets are required in all new homes. Industries are encouraged to cut water use through drought-tolerant landscaping and reuse of manufacturing water. There is strong evidence that such tactics have indeed reduced water use. A widely reported fact released in 1999 found that Los Angeles is using the same amount of water today as in 1972 despite the addition of 1 million people.

This new ethic to use water more efficiently addresses not only conservation, but water recycling – reusing treated wastewater. By recycling water and using it for other purposes such as landscape irrigation, regions that import water from distant sources can create a local, drought-resistant supply. Such projects have an added benefit of reducing waste-water discharge.

Yet water recycling is not without its drawbacks. There is the cost associated with construc-tion of separate treatment and delivery facilities in order to keep recycled water separate from potable supplies. Water quality can become an issue; each time water is used, its quality deteriorates. And public perception can become an obstacle when people consider the so-called yuck factor.

While some argue that efficiency measures such as these, along with water marketing, are sufficient to close the gap between supply and demand, others say that additional storage also is needed to capture flows in high water years such as 1997.

Water Marketing

Water marketing – the sale, ex-change or lease of water from one user to another – has the potential for becoming a key tool to meet increasing demands for water. Water transfers, however, raise a host of issues because of the unique nature of water, the interdependence of many users and the traditional use of the resource.

Some 80 percent of the so-called developed water is used by agriculture, a legacy of California's past. Demands for the future, however, are centered in the state's growing cities and the natural environment. To alleviate conflict over reallocation of some farm water to these other uses, many experts advocate a voluntary market method in which cities pay farmers for a portion of their water.

Such transfers, however, raise concerns about third-party impacts on those in the rural communities from farm workers to farm equipment dealers. Transferring water from farms to cities is an emotionally laden issue because whoever controls a region's water controls its destiny. Opponents of such transfers often refer to the Owens Valley – the first and most legendary transfer – as a source of their fears.

It was in the Owens Valley where Los Angeles purchased thousands of acres of land solely for the purpose of exporting the water via the Los Angeles Aqueduct. The aqueduct transformed what was once a community of small farms into a region of vast undeveloped range-land heavily dependent on tourism and recreation.

The 1987-1992 drought brought water transfers to the forefront.

Of necessity, water agencies arranged many transfers during the drought through a state-managed drought water bank. Even as interest in transfers increases and legislation is enacted to facilitate them, water marketing remains a hotly debated issue. ●

Saving the Salmon

Since the late 1980s, billions of dollars and millions of acre-feet of water have been expended in an unprecedented effort to restore salmon populations. Gravel for spawning beds has been deposited in waterways. Screens have been installed on diversion pipes.

Instream flow has been increased. Several small dams have been removed, with others slated for demolition.

Some of the most extensive efforts to develop an effective solution have occurred at Red Bluff Diversion Dam, a feature of the Central Valley Project. Completed in 1964, the dam spans the entire width of the Sacramento River southeast of Red Bluff and diverts water to local homes, three wildlife refuges and farms on the west side of the Sacramento Valley.

Although the dam has fish ladders on either side to allow for upstream migration, research revealed that they were inadequate. The dam gates are now left open most of the year.

Thirty-two state-of-the art fish screens were installed at the dam during a $17 million renovation project to keep fish out of a major irrigation canal. For two other canals, biologists are testing an Archimedes' screw type of pilot plant. As water is pumped into the canal, young salmon are passed through the plant and examined at a separate collection facility. The goal is to find the best way to not only keep fish out of the canals, but return them to the main river with the least amount of stress.

A view from inside the Archimedes' screw pilot pumping plant.

The San Joaquin River

A legal standoff over instream flows in the San Joaquin River gave way to consensus — at least temporarily — in 1999 as a pilot project sent water down a 15-mile stretch of the river that had been dry for some forty years.

Prior to construction of Friant Dam in the 1940s, water in the San Joaquin flowed unimpeded from its headwaters in the Sierra Nevada northwest along the valley floor to its rendezvous with the Sacramento River at the Delta. With the dam's completion, most of the San Joaquin's flow was diverted to farms on the east side of the San Joaquin Valley. Only a small flow of water is released below the dam, and the riverbed is usually dry from Gravelly Ford to Mendota Pool. (At Mendota Pool, water exported from the Delta is returned to the riverbed.)

Some 300,000 chinook salmon once swam up the San Joaquin River and its tributaries to spawn each year, and the San Joaquin supported both spring- and fall-run chinook salmon.

The spring-run became extinct in the early 1950s after Friant Dam began full operation. Its numbers were already depleted from years of river diversions and overfishing; state fishery biologists attempted to relocate the spring-run to the Merced River, but failed. Lack of fish flow releases from the dam also eliminated the fall-run chinook on the main river, although it continues to spawn in low numbers in San Joaquin River tributaries.

For years, fishery and environmental groups have pushed to return water to that stretch of the San Joaquin. In 1988, they sued, demanding that Friant Dam be compelled to release enough water to restore the salmon fishery. The organizations also pressed hard through the CVP Improvement Act to obtain fish flows from the dam. Federal officials backed away from even studying such an idea in the face of enormous opposition from farmers.

Positions hardened as both sides pressed on in court.

Flash forward eleven years. With the support of CalFed, the state-federal coalition seeking collaborative solutions on Delta issues, the divergent interest groups agreed on the pilot flow program.

Beginning on July 3, 1999, 35,000 acre-feet of water was gradually released from Friant Dam to help regenerate willow and cottonwood trees along the San Joaquin River. Environmentalists hope the flows will nourish the young riparian forest established during recent flood years and provide data on whether such flows impact neighboring water users.

The four-month pilot project offered little water supply risk for Friant water users. Some 20,000 acre-feet of the flow was to be delivered to them through another source and with $2.5 million from CalFed, they were prepared to purchase from other users the 12,000 acre-feet projected to be lost to recharge and evaporation between the dam and Mendota Pool. •

River Restoration

The development of much of California would have been impossible if not for the development of a system to move water. Development was favored, and control of water to benefit and protect humans was the goal.

Today's focus on balance has generated a new era in which rivers are being restored for the benefit of nature.

While it is impossible to turn back the clock and return to the past, river by river, stream by stream, creek by creek, Californians are working to restore pieces of the natural state. In a policy arena once dominated by single-minded pursuit of energy production, flood control or water supplies to benefit one area, decision makers now look at all the potential impacts on the entire system.

Known as watershed management, this movement attempts to bring together all the users of a particular waterway, upstream and downstream, to coordinate on solutions.

The creation of watershed conservancy groups on waterways both big and small throughout California stems not just from mandates at the federal and state levels but interest at the grassroots level by people whose yards back up to these streams.

At the state level, voters demonstrated their interest in environmental restoration when they approved $3 billion dollars in bonds in 1996 and 2000 to finance a host of projects designed to protect salmon, restore wetlands and improve water quality. •

originally altered for the good of man

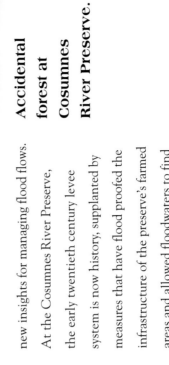

Accidental forest at Cosumnes River Preserve.

The Cosumnes River

For much of the past, environmental protection measures concentrated on mitigating damage caused by man, preserving what was left of the natural environment and stopping harmful practices. To that end, fish hatcheries were built to mitigate for lost habitat, scenic river stretches were safeguarded and obvious pollution sources were regulated.

These actions were generated by legislation and court battles. The restoration movement of the new century is based as much on science as politics. Through research, scientists have learned that the restoration of natural processes – not just mitigation – is necessary for healthy waterways, ecosystems and populations of flora and fauna.

At the Cosumnes River Preserve, a re-creation of the valley's inland sea demonstrated that floods are essential for the health of a river. In 1995, Nature Conservancy scientists at the 10,000-acre preserve on the outskirts

of Sacramento purposely breached a levee, flooding an area reclaimed for farming in the early 1900s. As water from the Cosumnes River spread across the field, it deposited seeds and silt. Within months, a young cottonwood and willow forest had taken root. Eventually, these cottonwoods will give way to valley oak, restoring a slice of the valley's natural landscape.

There were plans to breach another levee, but nature took its course during the 1997 floods. As the undammed Cosumnes River swelled to a record 90,000 cubic feet per second, the force of the water tore a hole in another preserve levee, flooding much of the land but causing only limited damage to the preserve's buildings. It was one of twenty breaks in the small, private farm levees that line the Cosumnes River.

As with other levees along California's rivers, those on the Cosumnes were built close to the river's edge, allowing more land to

be reclaimed for farming. However, these levees restrict the river's natural tendency to form bends and curves. As a result, when the river tries to change course during periods of high flow, it breaks a levee.

The 1997 floods brought a new respect for this natural process and

new insights for managing flood flows. At the Cosumnes River Preserve, the early twentieth century levee system is now history, supplanted by measures that have flood proofed the infrastructure of the preserve's farmed areas and allowed floodwaters to find their natural place. •

"To spend childhood days along creek banks is to be drawn into the wider world. A creek reaches upward into the hills and mountains, where clouds brood and gods bluster. It reaches down to the lowlands and the fat old rivers, sad and murky with the silt of experience. A creek teaches one the curve of the Earth, the youthful swell of mountains, the age of the seas.

Above all, a creek offers the mind a chance to penetrate the alien world of water and think like a tadpole or a trout. That is one of the great experiences of otherness, one of the leaps of perception that makes us human and allows us to live with dream and obscurity. What drifts in creek water is the possibility of other worlds inside and above our own.

Poet Robert Frost wrote, 'It flows between us, over us, and with us. And it is time, strength, tone, light, life and love.'"

– Peter Steinhart –
"The Meaning of Creeks"
Audubon magazine

Concluding Thoughts

Californians invented the modern state through water. They moved water from its origins to gold mines, farm fields and distant cities. They used the power of water to uncover gold ore, produce crops, generate electricity, indeed, to manufacture industrial prosperity. The California that we know today owes its existence to the development of water as a resource.

Yet this development came with an environmental price. Strip mining, the destruction of wetlands, the pollution of groundwater and surface water and the severe disturbance of natural habitat and species each contributed to the toll on nature caused by our actions. Beginning in the 1960s, the environmental consequences of such actions dominated newspaper headlines and public policy debates.

The role of technology cannot be overlooked in how water has shaped California. From the mile upon mile of wooden flumes built to mine gold to the ever more sophisticated levees and dams engineered to manage floods to the treatment plants constructed to disinfect drinking water and clean up wastewater, technology has advanced the effort to reshape the state.

Technology will no doubt play an important role in the future of California water. A focus on developing, among others, new appliances that use less water to clean clothes and dishes, new treatment techniques to clean up contaminated groundwater, new screens to keep fish out of harm's way will all assist in the twenty-first century goal of balance.

This year, 2000, marks the beginning of a new millennium. It also marks California's 150th anniversary. It is a year for reflection. A year in which not only the facts of the California water story deserve attention, but also the intangible way in which water touches our soul. Whether through a photograph or a verse, the art of water is as important as the science of water.

"Dams are not America's answer to the pyramids of Egypt. We did not build them for religious purposes and they do not consecrate our values (even if some are named after presidents).

Dams do, in fact, outlive their function. When they do, some should go."

— *Bruce Babbitt*
in an August 1998 speech

A New Ceremony for a New Age

Unlike the dam-related ceremonies of the past, the 1998 event at a small dam in the Sacramento Valley celebrated the structure's demolition, not its completion. In place of a ribbon cutting, this occasion featured a symbolic swing at the future as Interior Secretary Bruce Babbitt took a sledgehammer to McPherrin Dam.

The removal of McPherrin and three other dams on Butte Creek opened up 18.5 miles of spawning area to the spring-run chinook salmon. It was one of the most dramatic chapters in the ongoing restoration effort now underway to save populations of endangered salmon throughout California and the Pacific Northwest.

The dams were replaced with a screened, inverted siphon that draws water from the creek for use on the thousands of acres of rice fields within the boundaries of Western Canal Water District.

The Butte Creek program is considered a model for the new cooperative effort to restore the environment while protecting economic interests. •

The essential nature of water to sustain life and the environment will not change. Neither is it likely that the vagaries of the weather cycle will change. Water remains at once the least expensive item and most valuable element necessary to sustain life.

We are beyond the question of whether such development was good or bad, whether such development should or should not have occurred. The time is now to look to the future and use technology to sustain our cities and farms in the most environmentally sound way, and restore pieces of the natural state that was.

Technology alone is not the answer. In some places, the social cost may be too great as we try to turn the clock back and restore the natural state. That is something the next generation will have to determine, just as the past generation made its decisions.

The story of California water will continue to unfold over this century. For the foreseeable future, the next chapter will be devoted to finding equilibrium between the environment and the economy.

Achieving a balance is not easy. The scales tip first to one side. Then the other. Some would say it is an impossible dream. They may be right. But at this point in California's development, it is a scenario that offers hope. •

"Children of a culture born in a water-rich environment, we have never really learned how important water is to us. We understand it, but we do not respect it."

— *William Ashworth —*
Nor Any Drop to Drink

Timeline

1769 First permanent Spanish settlements established. Water rights established by Spanish law.

1772 First recorded sighting of the Delta by Spanish explorers.

1781 Dam erected on Los Angeles River when the Pueblo de los Angeles is founded.

1848 Treaty of Guadalupe Hidalgo signed, ceding California from Mexico to the United States. All property rights under Mexican law are protected, including private riparian and pueblo water rights. Gold discovered on the American River.

1849 California Gold Rush begins; miners develop system of appropriative rights by posting notice to hold water claims.

1850 California becomes the thirty-first state and adopts English common law, including its doctrine of riparian rights, in addition to the existing miner appropriative rights and pueblo rights systems. Office of Surveyor General established and charged with planning water projects. Flooding covers Sacramento's streets.

1859 Oliver Wozencraft promotes idea of irrigating Imperial Valley.

1860 State legislature authorizes formation of levee and reclamation districts.

1862 Homestead Act signed; increased numbers of settlers come to California and stake water claims. Great Calamity flood causes widespread flooding throughout the state.

1869-71 John Wesley Powell explores the Colorado River by boat; issues report of his findings in 1875.

1880 First flood control plan for the Sacramento Valley developed by State Engineer William Hammond Hall.

1884 Federal circuit court decision in *Woodruff v. North Bloomfield* ends hydraulic mining.

1886 California Supreme Court decision in *Lux v. Haggin* reaffirms legal preeminence of riparian rights; upheld again forty years later.

1887 Irrigation District Law, commonly referred to as the Wright Act, passed, permitting formation of irrigation districts.

1892 Conservationist John Muir founds the Sierra Club.

1901 First deliveries of water from the Colorado River to farmland in the Imperial Valley.

1902 U.S. Bureau of Reclamation established with passage of Reclamation Act.

1905 First bond issue for the city of Los Angeles' Owens Valley project; construction funded by second bond issue in 1907.

1905-07 Colorado River breaks through Imperial Valley Canal, creating Salton Sea.

1908 City of San Francisco's filings for Hetch Hetchy project approved.

1910-20 Groundwater pumping with diesel motors increases in the San Joaquin Valley; 600 wells in 1906, 5,000 in 1910, 11,000 in 1920.

1911 Imperial Irrigation District (IID) formed.

1913 Los Angeles Department of Power and Water (LADWP) completes Los Angeles Aqueduct, transporting surface water from the Owens Valley to the city of Los Angeles. Water Commission Act becomes effective; post-1914 appropriators required to comply with state's centralized water use permit process.

1914 Col. Robert B. Marshall of the U.S. Geological Survey proposes statewide plan for water storage and conveyance.

1919 State legislature authorizes officials to conduct statewide water resources investigation; report published in 1923.

1921 Colorado River Compact divides water between Upper and Lower Basins.

1922 Hetch Hetchy Valley flooded to produce water supply for San Francisco despite years of protest by John Muir and other conservationists.

1923 East Bay Municipal Utility District formed.

1926 California Supreme Court holds in *Herminghaus v. Southern California Edison* (200 Cal. 81) that a downstream riparian user can command the entire flow of a river to flood-irrigate land.

1928 In response to public outcry over *Herminghaus* decision, California Constitution is amended to include Article X, Section 2, which requires all water uses, not just appropriative, be reasonable and beneficial. Metropolitan Water District of Southern California (MWD) formed. Congress passes the Boulder Canyon Act, authorizing construction of Boulder (Hoover) Dam and other facilities on the Colorado River.

1928-34 Worst drought of the twentieth century in California; period later used as a measure for the storage and transfer capacity of all major water projects.

1929 East Bay Municipal Utility District completes Mokelume River Aqueduct.

1931 State Water Plan published, outlining utilization of water resources on a statewide basis. County of Origin Law passed, guaranteeing counties the right to reclaim their water from an exporter if it is ever needed in the area of origin. California Colorado River users enter into the Seven Party Agreement.

1933 Central Valley Project (CVP) Act passed by state Legislature.

1934 Construction starts on All-American Canal in the Imperial Valley and on Parker Dam on the Colorado River.

1937 Passage of the federal Rivers and Harbors Act of 1937 authorizes construction of the initial features of the CVP by U.S. Army Corps of Engineers.

1938 Construction begins on Shasta Dam, completed in 1945.

1940 Bureau completes All-American Canal and IID receives initial water deliveries.

1941 MWD completes the 242-mile long Colorado River Aqueduct.

1944 Mexican-American Treaty guarantees Mexico 1.5 million acre-feet annually from the Colorado River. Flood Control Act authorizes Corps to build a dam on the lower

Federal government assumes most costs of the Sacramento Valley Flood Control System with passage of Rivers and Harbors Act.

St. Francis Dam collapses and floods the Santa Clarita Valley, killing more than 450 people.

American River; Folsom Dam completed in 1956.

1945 Construction begins on Friant Dam on the San Joaquin River.

1951 State authorizes Feather River Project Act, later to become the State Water Project (SWP). First deliveries from the San Joaquin Valley.

1955 Flood in the Sacramento Valley.

1956 State Department of Water Resources established.

1959 Delta Protection Act enacted to resolve some issues of legal boundaries, salinity control and water exportation.

1960 Burns-Porter Act ratified by voters: $1.75 billion bond measure to build State Water Project.

1963 *Arizona v. California* lawsuit decided by the U.S. Supreme Court in Arizona's favor, allocating 2.8 million acre-feet of Colorado River water per year to Arizona.

1964 Partially completed Oroville Dam helps save Sacramento Valley from flooding.

1966 Construction begins on New Melones Dam on the Stanislaus River; completed in 1978.

1967 Legislature creates State Water Resources Control Board (State Board). Construction begins on Auburn Dam.

1968 Congress passes the Wild and Scenic Rivers Act. Oroville Dam completed.

1969 Congress passes National Environmental Quality Act.

1970 Passage of the California Environmental Quality Act and the California Endangered Species Act. First Earth Day held April 22. Los Angeles completes second aqueduct to transport surface water and groundwater from Owens Valley and Mono Lake to Los Angeles.

1972 California legislature passes own Wild and Scenic Rivers Act to preserve the north coast's remaining free-flowing rivers from development. Federal Clean Water Act passed to clean up the nation's polluted waters. First SWP deliveries to southern California.

1974 Congress passes Safe Drinking Water Act.

1975 Construction on Auburn Dam suspended for seismic investigations.

1976-77 Drought. 1977 is driest year on record.

1978 State Board issues Water Rights Decision 1485, setting Delta water quality standards.

1980 State-designated wild and scenic rivers added to federal Wild and Scenic Rivers Act.

1982 Proposition 9 (SB 200), the Peripheral Canal package, overwhelmingly defeated in a statewide referendum.

1983 California Supreme Court in *National Audubon Society v. Superior Court* rules that the public trust doctrine applies to Los Angeles' diversion from tributary streams of Mono Lake. Dead and deformed waterfowl discovered at Kesterson Reservoir, alerting people to problems of selenium-tainted agricultural drainage water.

1986 Ruling by the state court of appeals (Racanelli Decision) directs State Board to consider all beneficial uses, including instream needs, of Delta water when setting water quality standards. Passage of the Safe Drinking Water and Toxic Enforcement Act (Proposition 65) prohibiting the discharge of toxic chemicals into state waters.

Severe flooding nearly tops levees in Sacramento.

1987 State Board's Bay-Delta Proceedings begin to revise D-1485 water quality standards.

1987-92 Drought.

1989 Natural Resources Defense Council files suit to block renewal of CVP's Friant-Kern contracts. In a separate challenge to Los Angeles' Mono Basin water rights, an appellate court holds that fish are a public trust resource in *California Trout v. State Water Resources Control Board*. MWD and Imperial Irrigation District agree that MWD will pay for agricultural water conservation projects and receive the water conserved. Sacramento River winter-run chinook salmon declared a state endangered and federal threatened species.

1991 Memorandum of Understanding signed to implement urban water conservation programs. Inyo County and the city of Los Angeles agree to jointly manage Owens Valley water ending nineteen years of litigation. The West Coast's first municipal seawater desalination plant opens on Catalina Island. Emergency legislation is passed creating Drought Water Bank, allowing "buy back" of supplies while protecting water rights. The legislation is made permanent in 1992, allowing the drought water bank to occur in any year necessary.

1992 Congress approves landmark CVP Improvement Act.

1993 The federal court for the Eastern District in California in *Natural Resources Defense Council v. Patterson* rules that the CVP must conform with state law requiring release of flows for fishery preservation below dams. Delta smelt declared a federal and state threatened species.

1994 State Board amends Los Angeles' water rights licenses to Mono Lake, ending fifteen-year battle over Mono Lake. LADWP is restricted from diverting water until the lake rises 20 feet. Bay-Delta Accord signed, setting interim water quality standards to protect Delta estuary.

1995 State Board adopts new water quality plan for the Delta and begins hearings on water rights. Controversial water transfer agreement between IID and the San Diego County Water Authority proposed.

1997 New Year's storms cause state's second most devastating flood of the century. SWP's Santa Barbara Aqueduct completed.

1998 CalFed Bay-Delta Program continues efforts to forge a long-term plan to "fix" the Delta. California Colorado River users struggle to develop a plan to reduce state's use of river water under pressure by Interior Department and other six Colorado River states.

1999 Splittail minnow and spring-run chinook salmon added to endangered species list.

2000 Voters approve $2 billion bond to finance water quality, flood protection and environmental restoration measures. CalFed to release long-term Delta plan.

Words for Moving Water

© 1999. Used with permission of Blue Bear Books and Charles J. Soderquist.

"How far up a creek does a brook begin?"
– John Stilgoe, 1994

"To me, good weather is rain!"
– Marc Reisner, UC Davis speech, 1998

Editor's Note: This is a partial list of terms from Charles Soderquist's "Words for Moving Water." Some terms have been cut because of space constraints. This is not a technical glossary. A technical glossary appears on page 162.

Moving water word rules:
1. These words are water nouns. Each describes a moving water (stream, creek) or sometimes a water specifically communicating with moving water (lake, sea).
2. Water in its frozen state (iceberg, hail) or water in the atmosphere (cloud, fog) are included if they move.
3. Dubious examples (and words we just could not part with!) are noted with an asterisk (*).
4. All definitions are from Webster's Third New International, Unabridged Dictionary unless otherwise noted.

anabranch: a diverging branch of a river which reenters the main stream or which loses itself in sandy soil.

aqua-: of water.

aquaeductus*: the right in law to conduct water over the land of another.

aquaehaustus*: the right in law to draw water from a well, spring, or stream on another's land.

aquaeinmittendae*: the right in law to throw water from one's window on a neighbor's building or land.

arroyo: a brook, creek, stream; watercourse.

back stream: an eddy

backwash: water or waves washed or thrown back (as by oars or a propeller)

backwater: a water turned back in its course by an obstruction, an opposing current, or the flow of the tide.

bayou: 1. a creek, secondary watercourse, or minor river that is tributary to another river or other body of water. 2. a sluggish stream that follows a tortuous course through alluvial lowlands, swamps, or plantations. 3. a clear brook or rivulet that rises in the hills especially of northern Arkansas or southern Missouri. 4. an intermittent, partly closed, or disused watercourse that is sluggish or stagnant.

billow: a large swelling wave of water, especially in the open sea.

bogan*: pokelogan.

boil: 1. a swirling upheaval of water, especially one at the surface of a river, a large spring, a pool below a dam, or the sea. 2. the swirl made by a fish moving at or near the surface, especially when feeding.

bore: a tidal flood that regularly or occasionally rushes with a roaring noise into certain rivers or narrow bays of peculiar configurations or location and proceeds in one or more waves that often present a very abrupt front of considerable height dangerous to shipping.

brook: 1. creek. 2. a stream larger than a rill but smaller than a river, always of fresh water (New England) (def: Stilgoe, Shallow Water Dictionary, p 12).

brooklet: a small brook.

burn: 1. a stream, brook, rivulet (British). 2. water (Scottish).

capillary water: water that remains in the soil after 'gravitational water' has drained out.

cascade: a fall of water over steeply slanting rocks, especially a small fall or one of a series.

cataract: 1. a waterfall, especially a great fall of water over a precipice. 2. steep rapids in a large river. 3. an overwhelming down-pour or rush; flood.

channel: the deeper part of a moving body of water where the main current flows or which affords the best passage.

chute: 1. a fall or rapid. 2. *a quick descent, steep channel, or narrow sloping passage by which water falls to a lower level.

creek: 1. a small inlet or bay narrower and extending further into the land than a cove (New England). 2. a saltwater estuary of a small river or stream emptying on a low coast or into the lower reaches of a wide river (New England). 3. a natural stream of water normally smaller than and often tributary to a river (South and West).

crick: creek.

cut: a notch, creek, channel or inlet made by excavation or worn by natural action.

cutoff: a new and relatively short channel formed when a stream cuts through the neck of an oxbow.

deadwater: the eddy under that part of a vessel between the bottom of the stern and the wing-transon and buttock (def: Richard Henry Dana, Jr., A Seaman's Friend, 1851

debouchment: the mouth or outlet, especially of a river.

deluge: 1. an overflow of the land by water. 2. a drenching rain.

ditch*: a natural or artificial, usually narrow, watercourse or waterway.

downpour: a heavy drenching.

drain*: a watercourse, especially when narrow.

drainage: excess water removed from land by means of surface or sub-surface conduits.

drizzle: a light rain of very small drops, specifically falling at a velocity of 144 feet per hour and 2.25 feet per second (see shower).

ebb: river bend in which the current flows backward.

ebb tide: the tide ebbing or at ebb; flowing to sea.

eddy: a current of water running contrary to the main current, especially one moving circularly.

ephemeral stream: a stream that flows briefly during and following a period of rainfall in the immediate locality.

estuary: a water passage where the tide meets the current of a stream.

fall: a precipitous descent of water.

feeder: a tributary

firth: the opening of a river into the sea.

fjord: a narrow generally deep inlet of the sea between high cliffs or steep slopes.

flow: 1. the characteristic of movement of water showing unbroken continuity. 2. an arm or basin of the sea.

flux: a running stream, a continued flow.

foam: a light whitish mass of fine bubbles that is formed in or on the surface of a liquid by agitation (as ocean waves) or fermentation or effervescence.

ford: a shallow and usually narrow body of a river or other body of water that may be crossed by man or animal by wading.

forebay: the discharging end of a pond or millrace.

free water: groundwater free to move in response to gravity.

freshet: a great rise or flood or overflowing of a stream caused by heavy rains or melted snow.

frith: the opening of a river into the sea.

froth: an aggregation of bubbles formed in or on a liquid caused by agitation or fermentation.

geyser: a spring that throws forth intermittently jets of heated water and steam.

Godwater: rain in the arid, western U.S. (def: Marc Reisner, Cadillac Desert, p.4).

graupel: granular snow pellets, called also "soft hail."

gully washer: an extremely heavy fall of rain, usually of short duration.

gut: a narrow sea passage (as a strait); a small creek or narrow waterway in a marsh or tidal flat.

headwater: the source and upper part of a stream.

hot spring: a spring with water above 98° F.

hydro-: water

hygro-: humidity, moisture

ice: water reduced to the solid state by cooling.

iceberg: a large mass of land ice broken from a glacier at the edge of a body of water that when afloat has only a small part above the surface and that in the ocean floats with the subsurface currents often to great distances.

lagoon: a shallow sound, channel, pond, or lake near or communicating with the sea.

lake: 1. a small stream or channel. 2. a considerable inland body of standing water, an expanded part of a river, a reservoir formed by a dam, or a lake basin intermittently or formerly covered by water.

lakelet: a little lake.

leachate: a liquid that percolates through soil or other medium.

loch: a bay or arm of the sea, especially when nearly landlocked (Scottish).

lough: a bay or inlet of the sea.

lock: the space of water between the piers of a bridge.

logan*: short for pokelogan.

maelstrom: a powerful often destructive water current that usually moves in a circular direction with extreme rapidity sucking in objects within a given radius.

meander: a turn or winding of a stream.

meltwater: water from melting ice or snow.

millpond: water produced by damming a stream to produce a head of water for operating a mill.

mizzle: a fine rain.

outlet: 1. a stream flowing out of a lake or pond. 2. the lower end of a watercourse where its water flows into a lake or the sea.

percolate: a liquid that passes through a permeable substance.

planter*: a snag fixed at one end in a riverbed and standing almost rigidly (compare: sawyer).

pokelogan*: a usually stagnant inlet or marshy place branching off from a stream or lake.

pond: a body of water usually smaller than a lake and larger than a pool.

pondlet: a small pond.

pool: 1. a small and rather deep body of fresh water. 2. a quiet place in a stream.

pothole: a circumscribed body of water frequented by waterfowl.

puddle: a little pool of any kind.

rain: water falling in drops condensed from vapor in the atmosphere.

rapid: a part of a river where the current moves with great swiftness and the surface is usually is broken by obstructions but has no actual waterfall or cascade.

reach: 1. a straight portion of a river. 2. a level stretch of water between locks of a canal. 3. an arm of the sea extending up into the land.

redd: the spawning ground or nest of various fishes, as the salmon or the trout.

riffle: 1. a shallow extending across the bed of a stream over which the water flows swiftly so that the water is broken in waves. 2. a patch of ripples or small waves (as caused by a light breeze) on an otherwise calm and unbroken surface of water.

rill: a very small brook.

rime*: an accumulation of granular ice tufts on the windward side of exposed objects slightly resembling hoarfrost but formed only from undercooled fog or cloud and always built out directly against the wind.

rindle: runnel

ripple: 1. a shallow stretch of running water in a stream roughened or broken by rocky or uneven bottom. 2. a small wave propagated by both surface tension and gravity (distinguished by gravity wave).

river: a natural surface stream of water of considerable volume and permanent or seasonal flow.

rivulet: a small stream.

roller: one of a series of long heavy waves that roll in upon a coast (as after a storm).

runnel: rivulet, brook, streamlet, rindle.

sawyer*: a tree fast in the bed of a stream with its branches projecting to the surface and bobbing up and down with the current (compare: planter).

sea purse: sea puss.

sea puss: 1. a dangerous swirling of undertow due to the combined effect of several breakers. 2. an undertow setting along shore.

serein: mist or fine rain falling from an apparently clear sky.

sheetflood: an expanse of moving water into which the transient streams of arid regions spread out as they issue from the mountains upon the plains.

shower: a fall of rain that is of short duration or rapidly varying intensity over a limited area with drops usually about 1/25 inch in diameter and velocity of from 10 to 25 feet per second (see drizzle).

slew: slough.

slough: 1. a side channel or inlet. 2. a sluggish channel, a backwater. 3. a creek in a marshland, tidal flat, or bottom land.

slue: slough.

snowmelt: water from melting of snow.

snowslide: an avalanche of snow.

spoon-drift: water swept from the tops of waves by the violence of the wind in a tempest, and driven along before it, covering the surface of the sea. (def: Richard Henry Dana, Jr., A Seaman's Friend, 1851).

spray: an occasional sprinkling dashed from the top of a wave by the wind, or by its striking an object. (def: Richard Henry Dana, Jr., A Seaman's Friend, 1851).

spring: an issue of water from the earth.

sprinkle: light rain in scattered drops.

steam: water in the state of vapor.

strait: a comparatively narrow passageway connecting two large bodies of water.

stream: a body of running water flowing in a channel of the ground, in a cavern below the surface, or beneath or in a glacier.

streamlet: a small stream.

surge: a great rolling swell of water.

swell: a long relatively low wave or unbroken series of such waves.

tarn: a small steep-banked mountain lake or pool specifically in a basin produced by glacial erosion or deposition.

tidal river: a river up the course of which the tides are noticeable for a considerable distance.

torrent: 1. a rushing stream of water. 2. a raging flood, a tumultuous outpouring.

trickle: a thin, slow stream.

underflow: water moving through a subsurface material.

undertow: the current below the surface that sets seaward or along the beach when waves are breaking upon the shore.

vortex: a rapidly spinning current of water.

watercourse: 1. a stream of water. 2. specifically, a natural stream arising in a watershed and not wholly dependent upon surface water in its own immediate vicinity, flowing in a definite course either along a bed between visible banks or through a definite depression in surrounding lands, having a definite, permanent or periodic supply of water and a perceptible current in a particular direction, and discharging at a fixed point into a body of still or flowing water or disappearing underground.

waterdrop: a drop of water.

waterdust: particles of water composing clouds or fog.

waterfall: a perpendicular or very steep descent of the water of a stream.

water fog: a fine spray of fog formed by sending one high pressure stream of water against another in the tip of a nozzle.

watersmeet: a meeting place of two rivers.

waterspout: 1. a slender funnel-shaped or tubular column of rapidly rotating cloud-filled wind usually extending from the underside of a cumulus or cumulonimbus cloud down to a cloud of spray torn up by the whirling winds from the surface of an ocean or lake, being either straight and vertical or inclined and tortuous as it moves along and consisting largely of water. 2. a torrential burst of rain.

water vapor: water in a vaporous form especially when below boiling temperature and diffused.

whirlpool (whirlpool): water moving rapidly in a circle so as to produce a depression or cavity in the center into which floating objects may be drawn.

white water: frothy water.

Technical Glossary

acre-foot – common measurement for water; 325,851 gallons, or enough water to cover 1 acre of land (about the size of a football field) 1 foot deep. An average California household uses between 1/2 and 1 acre-foot of water per year.

appropriative right – water right based on physical control over water, or on a permit or license for its beneficial use.

aquifer – geologic formation that stores, transmits and yields significant quantities of water to wells and springs.

Bay-Delta – interconnected estuary region composed of the San Francisco Bay and Sacramento-San Joaquin River Delta. The Delta serves as the hub of the state's north to south water delivery system and supports hundreds of fish, wildlife and plants.

conjunctive use – the planned and coordinated use of surface water and groundwater supplies to improve water supply reliability.

desalination – specific treatment processes to demineralize sea water or brackish (saline) water for reuse.

developed water – water that is controlled and managed (dammed, pumped, diverted, stored in reservoirs or channeled in aqueducts) for a variety of uses.

diversion – term that refers to water diverted from a water source.

ecosystem – community of living organisms and their interrelated physical and chemical environment.

fish screens – physical structures placed at water diversion facilities – i.e. pipelines and canals – to keep fish from getting pulled into the facility and dying there.

floodplain – normally dry land area susceptible to being inundated by water during times of high flow.

groundwater – water that has seeped beneath the earth's surface and is stored in the pores and spaces between alluvial materials (sand, gravel or clay).

hydroelectric power – electricity produced by water-powered turbine generators.

hydrologic cycle – the natural recycling process powered by the sun that causes water to evaporate into the atmosphere, condense, and return to the earth as precipitation.

instream use – use of water within a river or stream, such as providing habitat for aquatic life, sport fishing, river rafting or scenic beauty.

irrigation – controlled application of water to crops to supplement that supplied by nature.

levee – embankment or raised area that prevents water from moving from one place to another.

off-stream – term that refers to a storage reservoir designed to hold water that is not directly on a stream or river.

on-stream – term that refers to a dam built across a stream or river, impounding the water in a reservoir.

nonpoint source pollution – Water pollution caused by diffuse sources with no discernible district source point, often referred to as runoff or polluted runoff, from agriculture, urban areas, mining, construction sites and other sources.

point source pollution – water pollution with a distinct, identifiable source point, such as from a pipe or channel.

pollution – adverse and unreasonable impairment of the beneficial uses of water even though no actual health hazard is involved.

precipitation – water falling as rain, snow or hail from the atmosphere to Earth.

reclaim – term that refers to making wasteland (desert or swamp) areas capable of being cultivated or lived on by irrigation or filling.

riparian area – land area directly influenced by a body of water often characterized by visible vegetation. Stream banks, lake borders and marshes are typical wetland areas.

riparian right – a water right based on the ownership of land bordering a river or waterway.

salinization – condition in which the salt content of soil accumulates over time to above the normal level; this occurs in some parts of the world where water containing high salt concentration evaporates from irrigated farm land.

selenium – Naturally occurring inorganic element found primarily in soils, and to a lesser extent in water and air. Selenium is a necessary nutrient in very small amounts but can be toxic in high doses.

surface water – water that remains on the earth's surface, in rivers, streams, lakes, reservoirs or oceans.

tributary – stream that flows into a larger stream, river or other body of water.

wastewater – water that contains unwanted materials from homes, businesses or factories.

water – odorless, tasteless, colorless liquid made up of a combination of two hydrogen atoms and one oxygen atom.

water allocation – the process of measuring a specific amount of water devoted to a given process.

water marketing – the transfer, lease or sale of water or water rights from one user to another.

water recycling – the treatment and reuse of wastewater to produce water of suitable quality for additional use.

water right – legal right to use a specified amount of water for beneficial purposes.

water treatment – process by which impurities are removed from water.

water year – twelve-month period, usually October 1 through September 30, over which yearly hydrologic data is measured.

watershed – region or land area drained by a river, stream or reservoir. Also called a drainage basin.

wetland – land area where water saturation is the dominant factor determining the nature of the soil and types of plant and animal communities.

Permissions

Ashworth, William. From *Nor Any Drop to Drink*. New York: Simon & Schuster, © 1982.

Brewer, William H. From *Up and Down California in 1860-1864*. Edited by Francis Farquhar. Berkeley: University of California Press, ©1948 by The regents of the University of California.

Cooper, Edwin. From *Aqueduct Empire: A Guide to Water in California, Its Turbulent History and Its Management Today*. Spokane: The Arthur H. Clark Company, ©1968. Reprinted by permission from The Arthur H. Clark Company.

Crespi, Juan. From *Fray Juan Crespi, Missionary Explorer on the Pacific Coast. 1769-1774*. Edited by Herbert E. Bolton. Berkeley: University of California Press, ©1927 by The Regents of the University of California.

Davis, Margaret Leslie. From *Rivers in the Desert*. New York, HarperCollins Publishers, ©1993. Reprinted by permission of Richard Curtis Associates, Inc.

DeVoto, Bernard. From *Across the Wide Missouri*. New York: Houghton Mifflin Co. © 1947, © renewed 1975 by Avis DeVotc and Jospeh R. Porter. Reprinted with permission from Houghton Mifflin Co.

Didion, Joan. From *The White Album*. New York: Farrar, Straus & Giroux, ©1979.

Elasser, Albert and Heizer, Robert. From *Natural World of the California Indians*. Berkeley: University of California Press ©1980 by The Regents of the University of California.

Everson, William. From "San Joaquin." Santa Rosa: Black Sparrow Press, ©1997 by

Jude Everson and the William Everson Literary Estate. Reprinted from *The Residual Years: Poems 1934-1948* with permission of Black Sparrow Press.

Hundley, Norris. From *The Great Thirst: Californians and Water, 1770s-1990s*. University of California Press, ©1992 by The Regents of the University of California.

Kelley, Robert. From *Battling the Inland Sea*. Berkeley: University of California Press, ©1989 by The Regents cf the University of California.

Leopold, Aldo. From *A Sand County Almanac*. New York: Oxforc University Press Inc., ©1949, 1977 by Oxford University Press, Inc.

Lopez, Barry Holstun. From *River Notes: The Dance of the Herons*. New York: Sterling Lord Literistic, Inc., ©1576 by Barry Holstun Lopez. Reprinted by permission of Sterling Lord Literistic, Inc.

McPhee, John. From *Annals of the Former World (Assembling California)*. New York: Farrar, Straus & Giroux, ©1998.

Muir, John. From *John of the Mountains*. Edited by Linnie Marsh Wolfe. New York: Houghton Mifflin Co., © 1938. Reprinted with permission from Houghton Mifflin Co.

Olson, Sigurd F. From *Open Horizons*. New York: Alfred A. Knopf, Inc.. ©1969.

Palmer, Timothy. From *Endangered Rivers and the Conservation Movement*. Berkeley: University of California Press, ©1986. Reprinted with permission by Timothy Palmer.

Percy, Marge. From *Circles on the Water: selected poems of Marge Percy*. New York Knopf ©1982.

Reeves, Richard. From "Vulnerable", a column by Richard Reeves. Kansas City: ©Universal Press Syndicate.

Roethke, Theodore. From *The Collected Poems of Theodore Roethke*. New York: Doubleday, ©1952 by Beatrice Roethke, Administratrix of the Estate of Theodore Roethke. Reprinted with permission from Bantam Doubleday Dell Publishing Group, Inc.

Rogers, Will. From "The People's Jester," a Will Rogers Column, 1932, TM/©2000. Permission granted by The Rogers Company under license authorized by CMG Worldwide Inc.

Saroyan, William. From *Fresno Stories*. New York: New Directions Pub. Corp., ©1994. Reprinted with permission by the Trustees of the Leland Stanford Junior University.

Shimer, John A. From *This Sculptured Earth: The Landscape of America*. New York: Columbia University Press, ©1959.

Smiley, Jane. From *A Thousand Acres*. New York: Alfred A. Knopf, Inc., ©1991 by Jane Smiley. Reprinted with permiss on from Alfred A. Knopf, Inc.

Snyder, Gary. From *A Place in Space: Ethics, Aesthetics, and Watersheds*. Washington, D.C.: Counterpoint, ©1995.

Starr, Kevin. From *Americans and the California Dream, 1850-1915*. New York: Oxford University Press, ©1973.

Stegner, Wallace Earle. From *The American West as Living Space*. Ann Arbor: The University of Michigan Press, ©1987.

Stegner, Wallace Earle. From: *Where the Bluebird Sings to the Lemonade Springs: Living and Writing in the West*. New York: Random House, ©1992.

Steinbeck, John. From *East of Eden*. New York: Penguin Putnam, Inc., ©1952 by John Steinbeck, renewed ©1980 by Elaine Steinbeck. John Steinbeck IV and Thom Steinbeck. Used by permission of Viking Penguin, a division of Penguin Putnam, Inc.

Steinbeck, John. From *The Grapes of Wrath*. New York: Penguin Putnam, Inc. ©1939, renewed ©1967 by John Steinbeck. Used by permission of Viking Penguin, a division of Penguin Putnam, Inc.

Steinhart, Peter. From the essay *The Meaning of Creeks*. Audubon. May 1989.

West, Paul. From *Out of My Depths: A Swimmer in the Universe*. New York: Doubleday, ©1983 by Paul West.

Photo Credits

The Water Education Foundation thanks all of the individuals and institutions that provided us with photos.

Title Page: Roland Mills

Acknowledgments Page: Hans Doe, Hans and Rita Schmidt Sudman

Contents Page: Pyramid Lake, California Department of Water Resources

Foreword Page: Roland Mills; Kevin Starr, state of California

Introduction Page: Roland Mills

A Natural State

Page 2, Rick Rickman. Page 3, courtesy California State Library. Pages 4-5, natural waterscape map, Curtis Leipold. Page 6, California Department of Water Resources. Page 7, Rick Rickman. Page 8, redwood forest, California Department of Water Resources. Page 9, Klamath River at Pacific Ocean, California Department of Water Resources. Page 10, Lake Tahoe, Larry Prosor. Page 11, Yosemite Falls, Scott Maiden. Page 12, Mt. Tallac, Larry Prosor. Page 13, Ben Klaffke. Pages 14 and 15, Joshua trees, Mojave Desert, California Department of Water Resources. Page 16, Rick Rickman. Page 17, courtesy California State Library. Page 18, courtesy California State Library. Page 19, left, courtesy California State Library; right, Santa Barbara Mission today, Rita Schmidt Sudman. Page 20, courtesy California State Library. Page 21, Monterey coast, California Department of Water Resources. Page 22, Roland Mills. Page 23, Shasta Reservoir, Sue McClurg.

The Rush

Page 24, Feather River, California Department of Water Resources. Page 25, Arthur M. McCurdy, courtesy City of Sacramento Archives. Page 26, Sutter's Mill in 1851, Charles Nahl, courtesy California State Library. Page 27, left, Sutter's Mill today, Christine Schmidt; right, Roland Mills. Page 28, courtesy California State Library. Page 29, top, courtesy California State Library; bottom, J.A. Todd, photographer, courtesy California Historical Society, San Francisco. Page 30, left, Malakoff Diggins today, Sue McClurg; center, Bob Brewer; right, Malakoff Diggins in 1882, J.A. Todd, photographer, courtesy California Historical Society, San Francisco. Page 31, left, Water Resources. Page 31, right, white-water rafting, Rick Rickman.

Arteries of Commerce

Page 32, Sacramento River, California Department of Water Resources. Page 33, courtesy California State Library. Page 34, top, courtesy City of Sacramento Archives; bottom, C. W. Winterbottom, courtesy City of Sacramento Archives. Page 35, Delta King today, Christine Schmidt. Page 36, Robert Campbell, courtesy Port of Oakland. Page 37, left, Deep Water Ship Channel, State Water Resources Control Board; top right, Rick Rickman; bottom right, Rick Rickman. Pages 38 and 39, Memoir and Maps of California, 1851, courtesy The Bancroft Library.

The Inland Sea

Page 40, Suisun Marsh, California Department of Water Resources. Page 41, courtesy California State Library. Page 42, courtesy Water Resources Center Archives, U.C. Berkeley. Page 43, Tulare Lake in 1997, G. Donald Bain. Page 44, the tule breakers, courtesy Phillips Library. Page 45, Bill Sleuter.

Flood Fights

Page 46, California Department of Water Resources. Page 47, view of Sacramento as it appeared in 1850 flood, courtesy California State Library. Page 48, U.S. Bureau of Reclamation. Page 49, Feather River levee break, 1997 flood, California Department of Water Resources. Page 50, both, California Department of Water Resources. Page 51, left, Department of Water Resources; right, 1986 flood, California Department of Water Resources; right, Rick Rickman. Page 52, left, 1955 flood, California Department of Water Resources; right, 1995 flood, Sunrise Boulevard in Citrus Heights, California Department of Water Resources. Page 53, Oroville Dam, under construction, during high flows in 1966, California Department of Water Resources. Page 54, courtesy California State Library. Page 55, left, Sutter Bypass levee break, 1997 flood, California Department of Water Resources; right, U.S. Bureau of Reclamation. Page 56, courtesy Imperial Irrigation District.

Page 57, Salton Sea, Sue McClurg. Page 58, both, San Diego Historical Society. Page 59, left, U.S. Bureau of Reclamation; right, coast highway near Malibu, Rick Rickman. Page 60, San Joaquin River levee break, 1997 flood, California Department of Water Resources. Page 61, flooded Sutter Bypass, 1997 flood, California Department of Water Resources.

Drought

Page 62, California Department of Water Resources. Page 63, Rick Rickman. Page 64, Rick Rickman. Page 65, California Department of Water Resources. Page 66, California Department of Water Resources. Page 67, Shasta Reservoir, U.S. Bureau of Reclamation. Page 68, left, Shasta Reservoir, Rick Rickman; right, California Department of Water Resources. Page 69, all, Rick Rickman. Page 70, top, Folsom Lake, California Department of Water Resources; bottom, Coyote Lake in San Jose, Rick Rickman. Page 71, left, California Department of Water Resources; center, Rick Rickman; right, Rick Rickman. Page 72, both, California Department of Water Resources. Page 73, Photographer unknown.

The Great Projects

Page 74, courtesy The Bancroft Library. Page 75, Hetch Hetchy Reservoir, Scott Maiden. Page 76, top, workers inspecting the State Water Project penstocks in Tehachapi Mountains, Rick Rickman; bottom, Rick Rickman. Page 77, all, California Department of Water Resources. Page 78, Shasta Dam, Rick Rickman. Page 79, left, Shasta Dam, Sue McClurg; right, Shasta Dam site, U.S. Bureau of Reclamation. Page 80, Hoover Dam, Rita Schmidt Sudman. Page 82, left, John Wesley Powell, courtesy U.S. Geological Survey; right, Hoover Dam, U.S. Bureau of Reclamation. Page 83, Metropolitan Water District of Southern California. Page 84, California Aqueduct, Boyle Engineering. Page 85, Los Angeles Aqueduct under construction and at dedication ceremony, both, Los Angeles Department of Water and Power. Page 86, dedication of Folsom Dam, U.S. Bureau of Reclamation. Page 87, both Los Angeles Department of Water and Power. Page 88, San Luis Reservoir, California Department of Water Resources. Page 89, U.S. Bureau of Reclamation. Pages 90-91, water map, Blue Cat Studio. Page 92, Roland Mills. Page 93, Sue McClurg.

Fertile Valleys

Page 94, Rick Rickman. Page 96, left courtesy California State Library; right, Rick Rickman. Page 97, left, cotton, California Department of Water Resources; right, Central Valley near Los Banos, Patrick Knisely, courtesy Metropolitan Water District of Southern California. Page 98, California Department of Water Resources. Page 99, left, courtesy California State Library; right, California Department of Water Resources. Page 100, left, aerial view of Sacramento Valley rice field, U.S. Bureau of Reclamation; right, drip irrigation, California Department of Water Resources. Page 101. top left, flood irrigation, California Department of Water Resources; bottom left, Rick Rickman; right, Larry Cumpton. Page 102, left, California Department of Water Resources; top right, harvesting hay, courtesy California State Library; bottom right, Rick Rickman. Page 103, main, Rick Rickman; inset, top to bottom, Rick Rickman, California Department of Water Resources, Sue McClurg, Rita Schmidt Sudman. Page 104, left, Santa Clara Valley Water District; right, California Department of Water Resources. Page 105, U.S. Geological Survey. Page 106, both, Paul Ecke Ranch. Page 107, Salinas Valley agriculture, both, Sue McClurg. Page 108, Leo Hetzel, courtesy Imperial Irrigation District. Page 110, California Department of Water Resources. Page 111, courtesy California State Library.

Part of Our Everyday Lives

Page 112, both, Rick Rickman. Page 113, left, Silicon Valley, Rick Rickman; right, Southern California Edison. Page 114, left, photographer unknown; right, both, Rick Rickman. Page 115, top left, water quality lab, Metropolitan Water District of Southern California; bottom left, California Department of Water Resources; right, T.C. Boyd, Wood Engraver, courtesy the Bancroft Library. Page 116, top, Rick Rickman, middle; California Department of Water Resources; bottom, Rick Rickman. Page 117, San Diego freeway, Patrick Knisely, courtesy Metropolitan Water District of Southern California. Page 118, left, courtesy California State Library; right, Rick Rickman. Page 119, all, Rick Rickman. Page 120, left, Sutro Baths. courtesy California State Library, right, mister cools sunbathers in Palm Springs, Rick Rickman. Page 121, Rick Rickman. Page 122, left, Delta houseboat, Rita Schmidt Sudman; right, courtesy California State Library. Page 123, fishing near Sacramento, Rick Rickman. Page 124, left, Rick Rickman; middle, courtesy California Department of Parks and Recreation; right, June Hammond, City of Sacramento Archives. Page 125, wind surfing, California Department of Water Resources. Page 126, Rick Rickman.

A Shift in Philosophy

Page 129, Mono Lake, Rick Rickman. Page 130, left, courtesy California State Library; right, Mono Lake, Mono Lake Committee. Page 131, Mark Leder-Adams. Page 132, left, U.S. Bureau of Reclamation; right, Tuolumne River, Tim Palmer. Page 133, left, Smith River; right, Klamath River, both by Tim Palmer. Page 134, bottom, Metropolitan Water District of Southern California; top, the Delta, California Department of Water Resources. Page 135, Metropolitan Water District of Southern California. Page 136, left, U.S. Bureau of Reclamation; right, California Department of Water Resources. Page 137, air stripping towers, California Department of Water Resources. Page 138, left, fall-run chinook salmon jumps up fish ladder at Folsom Dam, Rick Rickman; right, fishermen line up below Folsom Dam to catch returning salmon, Rick Rickman. Page 139, courtesy California State Library. Page 140, all, Rita Schmidt Sudman. Page 141, California Department of Fish and Game. Page 142, San Joaquin River Parkway, Rita Schmidt Sudman. Page 143, Burney Creek, California Department of Water Resources.

The Balance

Page 144, San Luis Reservoir, California Department of Water Resources. Page 145, Montezuma Slough, California Department of Water Resources. Page 146, Rita Schmidt Sudman. Page 147, Riparian forest in the Sacramento Valley, Sue McClurg, Page 149, Rick Rickman. Page 150, groundbreaking for Diamond Valley Reservoir, Metropolitan Water District of Southern California. Page 152, Geoff Fricker. Page 153, San Joaquin River, Ben Klaffke. Page 154, Sulfur Creek, California Department of Water Resources. Page 155, Michael Eaton. Page 156, Judy Wheatley. Page 157, Interior Secretary Bruce Babbitt, Rita Schmidt Sudman. Page 168, McClure Lake, California Department of Water Resources.

Bibliography

Selected Bibliography

Books

Adams, Emma H., *To and Fro, Up and Down in Southern California*, Cincinnati: W.M.B.C. press, 1887.

Austin, Mary Hunter, *The Land of Little Rain*, New York, NY, Boston, Mass.: Houghton, Mifflin and Company, 1903.

Baldon, Cleo and Melchior, IB, *Reflections on the Pool*, New York: Rizzoli, 1997.

Carson, Rachel, *Silent Spring*, Cambridge, Mass.: Houghton Mifflin Co., 1962.

Caughey, John W., *California: A Remarkable State's History*, Englewood Cliffs, N.J.: Prentice-Hall Inc., 1970.

Clappe, Louise Amelia Knapp Smith, *The Shirley Letters from California Mines in 1851-52*, San Francisco: T.C. Russell, 1922

Clemens, Samuel Langhorne, *Roughing It*, New York, NY, London, England: Harper & Brothers, 1924.

Dana, Richard Henry, *Two Years Before the Mast*, New York: Airmont Publishing Co., 1965.

Davis, Margaret Leslie, *Rivers in the Desert: William Mulholland and the Inventing of Los Angeles*, New York: HarperCollins Publishers, 1993.

Dillon, Richard, *Delta Country*, Novato, Calif.: Presidio Press, 1982.

——— *Exploring the Mother Lode Country*, Pasadena, Calif.: The Ward Ritchie Press, 1974.

——— *Fool's Gold*, Santa Cruz, Calif.: Western Tanager Press, 1981.

Duffy, William, Jr., *The Sutter Basin and Its People*, Davis, Calif.: The Printer, 1972.

Farquhar, Francis, editor, *Up and Down California in 1860-1864: The Journal of William H. Brewer, Professor of Agriculture in the Sheffield Scientific School from 1864 to 1903*, Berkeley: University of California Press, 1966.

Garner, William Robert, *Letters from California, 1846-1847*, Berkeley: University of California Press, 1970.

Hart, John, *Storm Over Mono: The Mono Lake Battle and The California Water Future*, Berkeley: University of California Press, 1996.

Haslam, Gerald, *The Other California: The Great Central Valley in Life and Letters*, Reno, Nev.: University of Nevada Press, 1990.

Holliday, J.S., *The World Rushed In: The California Gold Rush Experience*, New York: Simon and Schuster, 1981.

——— *Rush for the Riches: Gold Fever and the Making of California*, Oakland, Calif.: Oakland Museum and University of California Press, 1999.

Houston, James, *Californians: Searching for the Golden State*, Santa Cruz, Calif.: Otter B. Books, 1992.

Hedrick, U.P., *A History of Horticulture in American to 1860*, New York: Oxford University Press, 1950.

Hundley, Norris, Jr., *The Great Thirst: Californians and Water, 1770s-1990s*, Berkeley: University of California, 1992.

Johnson, Stephen, *The Great Central Valley: California's Heartland*, Berkeley: University of California Press in Association with California Academy of Sciences, 1993.

Kahrl, William, editor, *The California Water Atlas*, Sacramento: State of California, 1979.

Kelley, Robert, *Battling the Inland Sea: American Political Culture, Public Policy & the Sacramento Valley, 1850-1986*, Berkeley: University of California Press, 1989.

King, Clarence, *Mountaineering in the Sierra Nevada*, New York, NY: C. Scribner's Sons, 1902.

Kinsey, Don J., *The Romance of Water and Power*, Los Angeles: Department of Water and Power, 1926.

Knoles, George, H., editor, *Essays and Assays: California History Reappraised*, San Francisco: California Historical Society, 1973.

Lufkin, Alan, editor, *California's Salmon and Steelhead: The Struggle to Restore an Imperiled Resource*, Berkeley: University of California Press, 1991.

Madgic, Bob, *Pursuing Wild Trout: A Journey in Wilderness Values*, Anderson, Calif.: River Bend Books, 1998.

McArthur, Seonaid, editor, *Water in the Santa Clara Valley: A History*, Cupertino, Calif.: California History Center De Anza College, 1981.

McLaughlin, Mark, *Sierra Stories: True Tales of Tahoe, Carnelian Bay, Calif.:, Mic Mac Publishing, 1997.

McPhee, John, *Assembling California*, New York: Farrar, Straus and Giroux, 1993.

McWilliams, Carey, *California: The Great Exception*, Westport, Conn.: Greenwood Press, 1971.

Mitchell, Annie R., *Land of the Tules*, Fresno, Calif.: Valley Publishers, 1972.

Muir, John, *The Mountains of California*, New York, NY: The Century Co., 1911.

Myers, William, editor, *Historic Civil Engineering Landmarks of San Francisco and Northern California*, The History and Heritage Committee, San Francisco Section, American Society of Civil Engineers, Pacific Gas and Electric Co., 1977.

Palmer, Tim, *Endangered Rivers and the Conservation Movement*, Berkeley: University of California Press, 1986.

Pitt, Leonard, *The Decline of the Californios: A Social History of the Spanish-Speaking Californians, 1846-1890*, Berkeley: University of California Press, 1966.

Pourade, Richard, *The Glory Years, Volume 4 in The History of San Diego*, San Diego: Union-Tribune Publishing Co., 1964.

——— *Gold in the Sun, Volume 5 in The History of San Diego*, San Diego: Union-Tribune Publishing Co., 1964.

Reisner, Marc, *Cadillac Desert: The American West and its Disappearing Water*, New York: Viking, 1986.

Schawz, Joel, *A Water Odyssey: The Story of the Metropolitan Water District of Southern California*, Los Angeles: Metropolitan Water District of Southern California, 1991.

Stegner, Wallace, *Where the Bluebird Sings to the Lemonade Springs*, New York: Penguin Books, 1993.

Thompson, John and Dutra, Edward A., *The Tule Breakers The Story of the California Dredge*, Stockton, Calif.: Stockton Corral of Westerners and University of the Pacific, 1983.

Treadwell, Edward T., *The Cattle King*, Santa Cruz, Calif.: Western Tanager Press, 1981.

Treutlein, Theodore, *San Francisco Bay, Discovery and Colonization, 1769-1776*, San Francisco: California Historical Society, 1968.

Turner, Tom, *Sierra Club: 100 Years of Protecting Nature*, New York: H.N. Abrams in association with the Sierra Club, 1991.

Vileisis, Ann, *Discovering the Unknown Landscape: A History of America's Wetlands*, Washington, D.C.: Island Press, 1997.

Walters, Shipley, *Clarksburg, Delta Community*, Yolo County Historical Society, 1988.

Wickson, E.J., *California Nurserymen and the Plant Industry, 1850-1910, A Series of Addresses*, Los Angeles: California Association of Nurserymen, 1921.

Reports

The Bay Institute of San Francisco, *From the Sierra to the Sea: The Ecological History of the San Francisco Bay-Delta Watershed*, San Rafael, Calif.: 1993

California Department of Water Resources, *The 1976-1977 California Drought: A Review*, Sacramento, Calif. 1978

——— *California State Water Project Atlas*, Sacramento, Calif.: 1999

——— *The California Water Plan Update, Bulletin 160-98*, Sacramento, Calif.: 1998

California Resources Agency, Final Report, *Governor's Flood Emergency Action Team*, Sacramento, Calif.: 1997

Los Angeles Chamber of Commerce, *What the Newcomer Should Know About Agriculture in Los Angeles County*, Los Angeles: 1941

U.S. Army Corps of Engineers, *Sacramento and San Joaquin River Basins, Post-Flood Assessment*, Sacramento, Calif.: 1999

U.S. Geological Survey, *National Water Summary 1988-89, Hydrologic Events and Floods and Droughts*, Denver, Colo.: 1991

Periodicals

Ashcroft, Lionel, "Mexican Rule" (1997), available www.marinweb.com Internet

Bokovoy, Matthew F., "Inventing Agriculture in Southern California," The Journal of San Diego History (Spring 1999), www.edweb.sdsu.edu/sdhs Internet

Christensen, Jon, "Flood of Values," Nature Conservancy (July/August 1997) 8-9

Eaton, Mike, "New Years Flood Creates Problems and Opportunities," The Nature Conservancy of Californ a Newsletter (Spring 1997) 1-3

Ferguson, Lilian, "Garden Swimming Pools", Sunset Magazine (February 1901)

Holder, Charles, "Rod, Reel and Gaff in Southern California," Sunset Magazine (January 1901) 73-83

Larkin, Edgar L., "Wonderful New Inland Sea," Cosmopolitan Magazine (October 1906)

London, Jack, "The Story of an Eyewitness," Collier's (May 5 1906)

Lund, LeVal, "Past, Present and Future of the Los Angeles Water System" Water World Magazine (1999)

MacDougal, D.T., "A Voyage Below Sea-Level On the Salton Sea," The Outing Magazine (February 1908)

Pattison, Kermit, "Why Did the Dam Burst?" Invention and Technology (Summer 1998) 23-31

Rolle, Andrew, "Turbulent Waters: Navigation and California's Southern Central Valley," California History (Summer 1996) 128-137

Roos, Maurice, "Drought and Water Management in California," Chapter 14 of The Arid Frontier, Netherlands: Kluwer Academic Publishers, 1998

Vallejo, Guadalupe, "Ranch and Mission Days in Alta California," The Century Magazine (December 1890), available www.sfmuseum.org Internet

Van Hoosear, J.E., "'Pacific Service' Supplies the World's Largest Baths," PG&E Magazine (September 1912), www.sfmuseum.org Internet

Wierzbicki, Felix Paul, "California As It is And As It May Be, Part V: The Natives of California," The Californians, (Volume 12, No. 6), 40-45

Woehlke, Walter W., "After the Great Drouth," Sunset Magazine (December 1924)

Woehlke, Walter, W., "The Land Before-and-After," Sunset Magazine (April 1912)